Bird-dropping
of the
British Isles

A Field Guide to the Bird-dropping Mimics

Jon Clifton
Jim Wheeler

www.norfolkmoths.co.uk

Anglian Lepidopterist Supplies
Specialising in moth traps and related equipment

www.angleps.com

1st Edition. 2011

Published by Clifton & Wheeler

Copyright © J. Clifton, J.R. Wheeler 2011

The authors assert their moral right to be identified as the authors of this work. All rights reserved. No part of this publication may be reproduced, stored in a retrieval system, or transmitted in any form or by any means, electronic, internet, mechanical, photocopying, recording or otherwise, without prior permission of the publisher and copyright holders.

Printed in the United Kingdom by Henry Ling Limited, at the Dorset Press, Dorchester, DT1 1HD

ISBN 978-0-9568352-0-8

Cover Photograph
Gypsonoma dealbana
by **Jim Wheeler**

Cover Illustration
Epiblema cynosbatella
by **Richard Lewington** 2011

Introduction

Many species of moths have evolved cryptic black and white markings resembling bird-droppings as a means of camouflage. The Chinese Character (*Cilix glaucata*) is a classic example and one that all moth recorders are familiar with. There are many such species within the family Tortricidae, especially in the subfamily *Olethreutinae*.

In this guide we aim to show classic examples of bird-dropping mimics, such as the *Apotomis*, *Hedya* and *Epinotia* and have also included some of the cryptic black and white *cochylids*. Unfortunately there is no definitive list of bird-dropping species and naturally, one recorders list will always differ from the next. To overcome this, a provisional list with recommendations was circulated to several experienced lepidopterists for selecting species for inclusion. We hope that most readers will agree with the final selected list.

We have decided to include species such as *Spilonota laricana* due to its close affinity with the dark form of *S. ocellana*. Furthermore, we are also including *Epiblema tetragonana*, *E. cnicicolana* and *E. sticticana* due to their similarity to *E. scutulana* and *E. cirsiana*.

The moths are shown in their resting position, alongside a set specimen, with a brief account giving distinguishing features and use of coloured pointers to highlight the critical areas in a user friendly way. Maps show Vice-county distribution across the British Isles, alongside seasonal flight graphs and life size representations of each species, to aid identification.

A good quality hand lens should be used for viewing as some of the markings of the smaller individuals are almost impossible to see with the naked eye. As with any moth, worn examples will always cause pitfalls and identification issues. If in doubt, retain the specimen.

With photography, lighting and camera settings can give a false impression. Some colours can be over or under-exposed, size and shape may be masked. Always retain sightings of scarce species until confirmed by an experienced recorder otherwise the record may not be accepted. Remember, some will require determination by genitalia examination.

Acknowledgements

We would like to thank Allan Drewitt, Neil Sherman, Mike Wall and all other photographers for allowing us to reproduce their images. Special thanks must go to Mark Parsons for comments and useful discussion in the preparation of this guide and Dr. John Langmaid for providing the Vice-county level data that has allowed us to database and map the species distribution across Great Britain and Ireland.

Please continue to send your moth records to your local County Moth Recorder.

Morphology

1 costa	**A** thorax	**F** apical spot/apical area
2 apex	**B** tegula	**G** subterminal fascia
3 termen	**C** basal and subasal fascia	**H** ocellus
4 tornus	**D** median fascia	**I** pre-tornal mark
5 dorsum	**E** strigulae	**J** cilia
		K dorsal patch/dorsal blotch

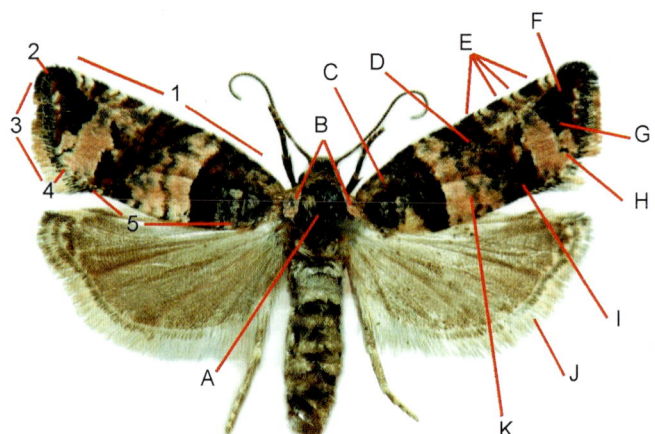

Adult flight times

Jan	Feb	Mar	Apr	May	Jun	Jul	Aug	Sep	Oct	Nov	Dec
			First Generation				Second Generation				

Flight graphs generated by digital distribution data and represent approximate flight periods based on the available data. **Flight times will vary geographically**. Species *National Status* is based on that generated by MapMate™ V.2.4. 2011 (patch 425).

All maps copyright © Clifton & Wheeler 2011. Underlying maps produced by MapMate™ using Digital Map Data © HarperCollins-Bartholomew 2007. Data Overlay © Jim Wheeler - NOLA Database System 2011.

Vice-county species distribution data (up to 2009) provided by Dr J.R. Langmaid, co-author of the *Microlepidoptera Reviews*, published annually in the *Entomologist's Record and Journal of Variation*, by J.R. Langmaid & M.R. Young.

All Photographs are the exclusive copyright of their respective owners.

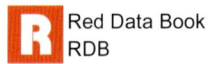 Red Data Book RDB

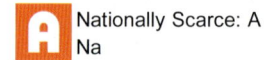 Nationally Scarce: A Na

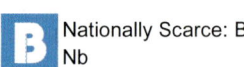 Nationally Scarce: B Nb

Contents

Species List

[0923]	Phtheochroa sodaliana	(Haworth, 1811)	Page 5
[0924]	Hysterophora maculosana	(Haworth, 1811)	Page 6
[0925]	Phtheochroa rugosana	(Hübner, 1799)	Page 7
[0926]	Phalonidia manniana	(Fischer von Röslerstamm, 1839)	Page 8
[0954]	Eupoecilia angustana	(Hübner, 1799)	Page 9
[0964]	Cochylis dubitana	(Hübner, 1799)	Page 10
[0964a]	Cochylis molliculana	(Zeller, 1874)	Page 11
[0965]	Cochylis hybridella	(Hübner, 1813)	Page 12
[0966]	Cochylis atricapitana	(Stephens, 1852)	Page 13
[0967]	Cochylis pallidana	(Zeller, 1847)	Page 14
[0968]	Cochylis nana	(Haworth, 1811)	Page 15
[1048]	Acleris variegana	([Denis & Schiffermüller], 1775)	Page 16
[1066]	Celypha woodiana	(Barrett, 1882)	Page 17
[1079]	Piniphila bifasciana	(Haworth, 1811)	Page 18
[1082]	Hedya pruniana	(Hübner, 1799)	Page 19
[1083]	Hedya nubiferana	(Haworth, 1811)	Page 20
[1084]	Hedya ochroleucana	(Frölich, 1828)	Page 21
[1086]	Hedya salicella	(Linnaeus, 1758)	Page 22
[1085]	Metendothenia atropunctana	(Zetterstedt, 1840)	Page 23
[1089]	Apotomis semifasciana	(Haworth, 1811)	Page 24
[1090]	Apotomis infida	(Heinrich, 1926)	Page 25
[1091]	Apotomis lineana	([Denis & Schiffermüller], 1775)	Page 26
[1092]	Apotomis turbidana	(Hübner, 1825)	Page 27
[1093]	Apotomis betuletana	(Haworth, 1811)	Page 28
[1094]	Apotomis capreana	(Hübner, 1817)	Page 29
[1095]	Apotomis sororculana	(Zetterstedt, 1839)	Page 30
[1096]	Apotomis sauciana	(Frölich, 1828)	Page 31
[1097]	Endothenia gentianaeana	(Hübner, 1799)	Page 32
[1098]	Endothenia oblongana	(Haworth, 1811)	Page 33
[1099]	Endothenia marginana	(Haworth, 1811)	Page 34
[1108]	Lobesia abscisana	(Doubleday, 1849)	Page 35

[1123]	Ancylis laetana	(Fabricius, 1775)	Page 36
[1132]	Epinotia subocellana	(Donovan, 1806)	Page 37
[1133]	Epinotia bilunana	(Haworth, 1811)	Page 38
[1134]	Epinotia ramella	(Linnaeus, 1758)	Page 39
[1135]	Epinotia demarniana	(Fischer von Röslerstamm, 1840)	Page 40
[1151]	Epinotia trigonella	(Linnaeus, 1758)	Page 41
[1167]	Gypsonoma aceriana	(Duponchel, 1843)	Page 42
[1168]	Gypsonoma sociana	(Haworth, 1811)	Page 43
[1169]	Gypsonoma dealbana	(Frölich, 1828)	Page 44
[1170]	Gypsonoma oppressana	(Treitschke, 1835)	Page 45
[1173]	Gibberifera simplana	(Fischer von Röslerstamm, 1836)	Page 46
[1174]	Epiblema cynosbatella	(Linnaeus, 1758)	Page 47
[1176]	Epiblema trimaculana	(Haworth, 1811)	Page 48
[1177]	Epiblema rosaecolana	(Doubleday, 1850)	Page 49
[1178]	Epiblema roborana	([Denis & Schiffermüller], 1775)	Page 50
[1179]	Epiblema incarnatana	(Hübner, 1800)	Page 51
[1180]	Epiblema tetragonana	(Stephens, 1834)	Page 52
[1184]	Epiblema scutulana	([Denis & Schiffermüller], 1775)	Page 53
[1184a]	Epiblema cirsiana	(Zeller, 1843)	Page 54
[1185]	Epiblema cnicicolana	(Zeller, 1847)	Page 55
[1186]	Epiblema sticticana	(Fabricius, 1794)	Page 56
[1187]	Epiblema costipunctana	(Haworth, 1811)	Page 57
[1197]	Eucosma campoliliana	([Denis & Schiffermüller], 1775)	Page 58
[1198]	Eucosma pauperana	(Duponchel, 1843)	Page 59
[1205]	Spilonota ocellana	([Denis & Schiffermüller], 1775)	Page 60
[1205a]	Spilonota laricana	(Heinemann, 1863)	Page 61
[1236]	Pammene fasciana	(Linnaeus, 1761)	Page 62
[1256]	Cydia servillana	(Duponchel, 1836)	Page 63
		Thumbnail Finder	Page 64
		Photographers	Page 68
		References	Page 69
		Index	Page 70

Tortricidae: Cochylinae
Phtheochroa sodaliana (Haworth, 1811) [0923]

Photo © Allan Drewitt

Photo © Jon Clifton

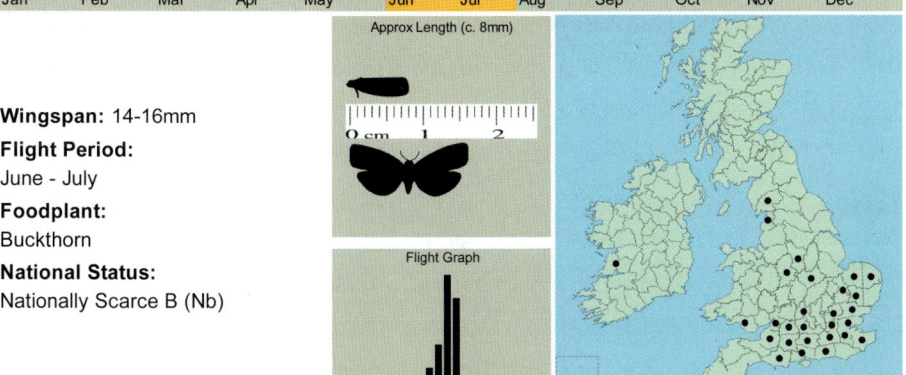

Wingspan: 14-16mm
Flight Period:
June - July
Foodplant:
Buckthorn
National Status:
Nationally Scarce B (Nb)

Larger than *Hysterophora maculosana* showing a whitish head and thorax (A). Forewing ground colour white with distinct black marks in the median fascia (B), tornal area showing a distinct black stria (C) with an obvious rufous apical spot (D). Hindwing mottled (E).

Mainly found in southern counties of England where it is very local, occurring on calcareous soils. Not recorded in Scotland and scarce in Ireland.

Adult moths can sometimes be found sitting on the leaves of the foodplant in the early evening, flying usually for a short period in late evening. The larvae feed on the berries of the foodplant in July and August, spinning them together.

Similar Species: *Hysterophora maculosana* Page 6

Tortricidae: Cochylinae

Hysterophora maculosana (Haworth, 1811) [0924]

Photo © Neil Sherman (Male)

Photo © Mike Wall (Brian Elliott Collection)

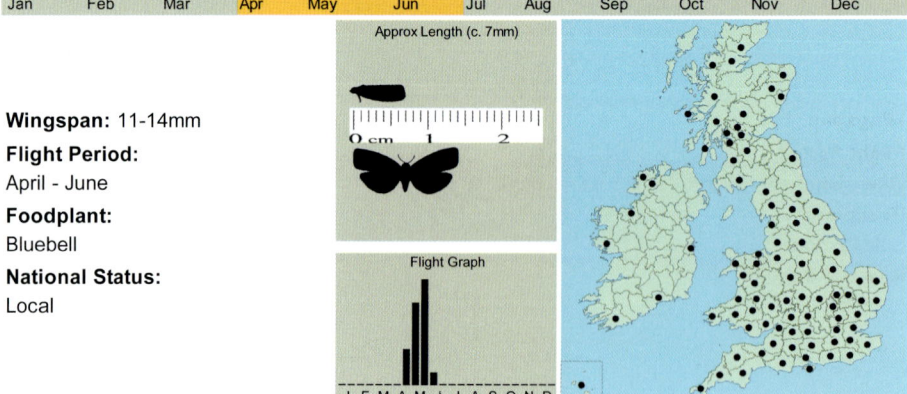

Wingspan: 11-14mm
Flight Period:
April - June
Foodplant:
Bluebell
National Status:
Local

Separated from *Phtheochroa sodaliana* by its smaller size and darker head (**A**). The median fascia mixed with black and grey markings centrally (**B**), apex of wing with a rich brown spot (**C**). A good hand lens is required to see this, caution is needed with worn examples.

Sexual dimorphism pronounced. The female (**D**) being much greyer overall with all markings becoming diffuse. The hindwing is completely greyish-brown lacking the white ground colour of the male.

Local in woodlands throughout much of the British Isles.

Adult males fly in sunshine around the foodplant. The larvae feed in the seed-heads of the foodplant in July and August.

Similar Species: *Phtheochroa sodaliana* Page 5

Tortricidae: Cochylinae

Phtheochroa rugosana (Hübner, 1799) [0925]

Photo © Allan Drewitt   Photo © Jon Clifton

Wingspan: 16-20mm
Flight Period:
May - July
Foodplant:
White Bryony
National Status:
Common

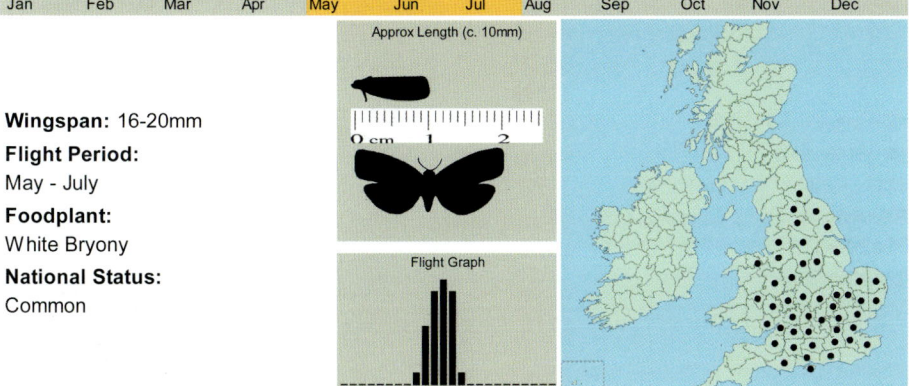

A large and distinctive species that should present little confusion. Forewing being rough scaled with tufts and raised scales present. Note the large white inward oblique marking at the median fascia (A).

Often seen flying in woodlands. Recorded in central and southern England as far north as County Durham.

Adult moths may be seen on warm evenings flying around the foodplant, but more active at dusk when they come to light. The larvae feed on the flowers, berries and stems of the foodplant from June to September.

Tortricidae: Cochylinae

Phalonidia manniana (Fischer von Röslerstamm, 1839) [0926]

Photo © Mike Crewe   Photo © Jon Clifton

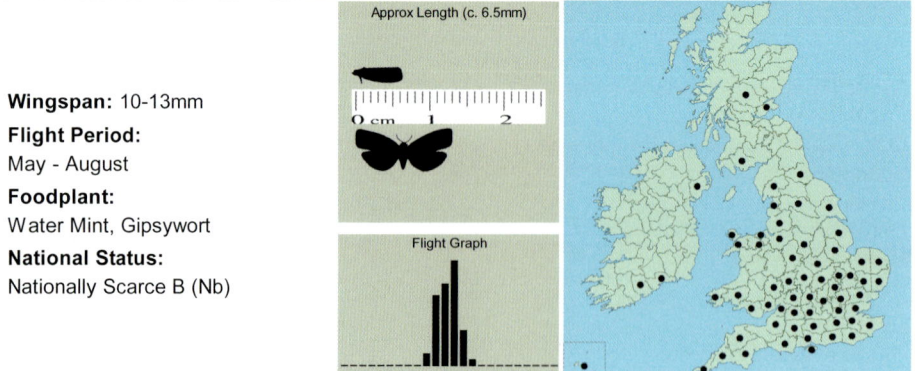

Wingspan: 10-13mm
Flight Period:
May - August
Foodplant:
Water Mint, Gipsywort
National Status:
Nationally Scarce B (Nb)

A fairly distinctive species showing a whitish ground colour to the forewing, a black dorsal part to the median fascia (A) (this area can appear broader in some individuals), and an ochreous pre-apical spot (B).

Found very locally in marshes, river-banks and other freshwater margins, predominately in southern England, but also scarcely recorded from Wales and northern England. Very few records from Scotland and Ireland.

Adult moths fly freely in the evening and come to light. The larvae feed in the stems of the foodplants from August to October.

Tortricidae: Cochylinae

Eupoecilia angustana (Hübner, 1799) [0954]

Photo © Allan Drewitt

Photo © Jon Clifton

Wingspan: 10-15mm

Flight Period:
May - September

Foodplant:
Plantains, Yarrow, Sneezewort, Marjoram, Goldenrod, thymes, Heather, Sitka Spruce

National Status:
Common

The warm ochreous markings of this species should immediately rule out the vaguely similar **Cochylis atricapitana** and **C. dubitana**. The basal and tornal areas showing diffuse warm ochreous markings (A), median fascia and terminal area black (B). Hindwing mottled grey (C).

The heathland form *fasciella* is slightly smaller than the nominate form, appearing whiter. The Shetland subspecies *thuleana* possess narrower wings with the forewing markings reduced to a suffusion of ochreous with a pale crescent shaped marking just beyond the median fascia. This being very rare in Shetland only forming a minority of the population found there, most records being of the nominate form and mainly confined to the northern isles.

The nominate form **angustana** occuring in woodland edges, meadows and similar locations. The form **fasciella** occuring on heathland and moorland throughout the British Isles. A third form **thuleana** (red map dot), considered a subspecies, is found on Shetland.

Adults fly from dusk, with the form *fasciella* flying in abundance on warm afternoons. The larvae feed on the flowers and seed-heads of various foodplants from July to October.

Tortricidae: Cochylinae
Cochylis dubitana (Hübner, 1799) [0964]

Photo © Neil Sherman   Photo © Jon Clifton

| Jan | Feb | Mar | Apr | May | Jun | Jul | Aug | Sep | Oct | Nov | Dec |

Wingspan: 12-14mm

Flight Period:
Two Generations
May - June, July - September

Foodplant:
Ragwort, hawk's-beard, hawkweed, Perennial Sow-thistle, Goldenrod and other Compositae

National Status:
Local

One of three similar looking *cochylids*. Separated from **Cochylis hybridella** by the lack of a white thorax, and from **C. atricapitana** by the white head and labial palpi (black in *C. atricapitana*) (**A**).

Locally widespread in the south of Britain, becoming scarce in the north. Often found on rough grassy hillsides, fens, railway banks and waste ground.

Adult moths can be easily disturbed in late afternoon, flying freely in the evening. The larvae feed inside the flowers and developing seed-heads of various Compositae in July, with those from the second generation in September.

Similar Species: *Cochylis atricapitana* Page 13   *Cochylis hybridella* Page 12

Tortricidae: Cochylinae
Cochylis molliculana (Zeller, 1874) [0964a]

Photo © Mike Wall   Photo © Graham Finch

| Jan | Feb | Mar | Apr | May | Jun | Jul | Aug | Sep | Oct | Nov | Dec |

Wingspan: 12-14mm
Flight Period:
June - August
Foodplant:
Bristly Oxtongue
National Status:
Unknown

Fairly distinctive. Ground colour of forewing suffused ochreous-cinnamon. The sub-basal, median and subterminal fascia consisting of a brown blotch on the costa (**A**), the median fascia displaced medially forming a rufous elongate patch on the dorsum (**B**). The ground colour and intensity of markings can vary.

First recorded in Britain at Portland, Dorset in 1993, this species has rapidly colonised the south coast of England, spreading inland as far north as Worcestershire and Huntingdonshire. The larvae feed within the seed-heads of the foodplant in the Autumn.

Similar Species: *Cochylis hybridella* Page **12**

Tortricidae: Cochylinae

Cochylis hybridella (Hübner, 1813) [0965]

Photo © Allan Drewitt　　　　　Photo © Jon Clifton

| Jan | Feb | Mar | Apr | May | Jun | Jul | Aug | Sep | Oct | Nov | Dec |

Wingspan: 12-15mm

Flight Period:
July - August

Foodplant:
Bristly Oxtongue, Hawkweed Oxtongue, hawk's-beard

National Status:
Local

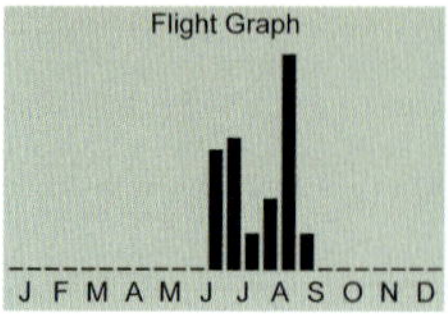

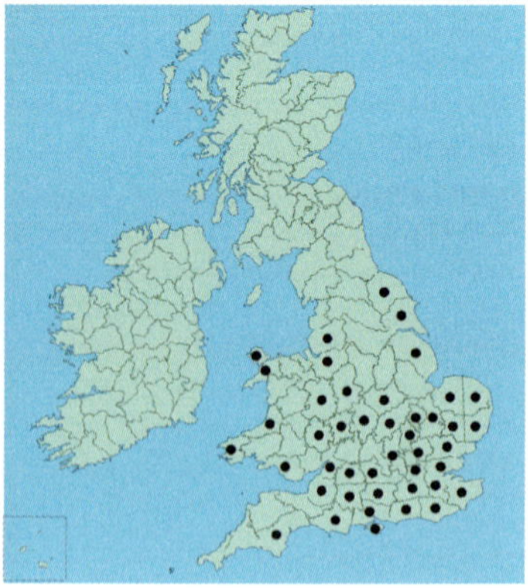

Superficially similar to **Cochylis atricapitana** or **C. dubitana** but immediately separated from those species by the white thorax (A).

Occurring mainly in chalk and limestone districts, rough grassland, scrub and coastal habitats. Locally recorded in southern England and Wales. Recorded as far north as Yorkshire but apparently rare.

Adult moths fly freely at sunset and come to light. The larvae feed in the seed-heads of the foodplant in August and September.

Similar Species:　Cochylis dubitana Page 10　　Cochylis atricapitana Page 13

Tortricidae: Cochylinae

Cochylis atricapitana (Stephens, 1852) [0966]

Photo © Allan Drewitt

Photo © Jon Clifton

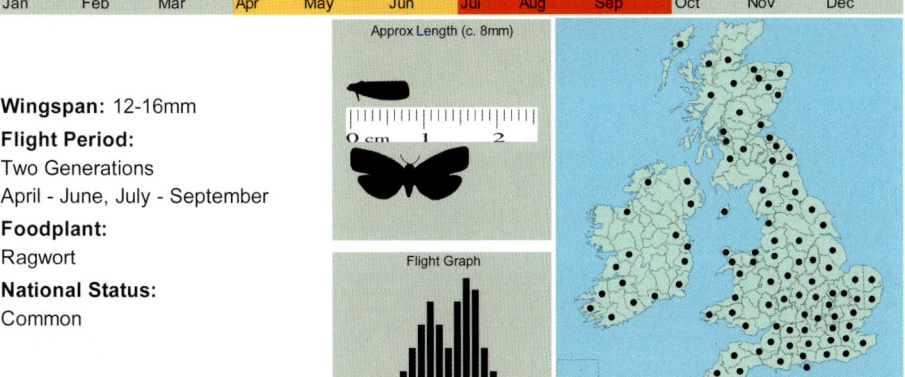

Wingspan: 12-16mm
Flight Period:
Two Generations
April - June, July - September
Foodplant:
Ragwort
National Status:
Common

Separated from *Cochylis hybridella* by the lack of a white thorax and from *C. dubitana* by the black head and labial palpi (A). Beware of worn specimens that could cause confusion.

Regularly recorded at coastal areas, and to a lesser extent, inland chalk downs and rough pasture land where the foodplant occurs.

Adult males fly at sunset, the female a little later, and both sexes may be taken at light. The larvae feed on the flowers and shoots of the foodplant in July and September.

Similar Species: *Cochylis dubitana* Page **10**   *Cochylis hybridella* Page **12**

Tortricidae: Cochylinae
Cochylis pallidana (Zeller, 1847) [0967]

Photo © Jim Wheeler

Photo © Mike Wall (John Langmaid Collection)

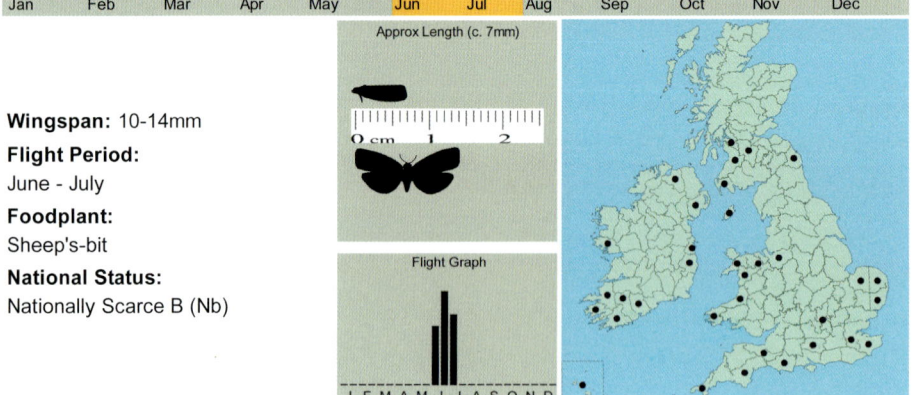

Wingspan: 10-14mm
Flight Period:
June - July
Foodplant:
Sheep's-bit
National Status:
Nationally Scarce B (Nb)

Similar to **Cochylis nana** but is larger and shows a narrower median fascia being separate from the basal fascia (A).

Scattered records from mainly coastal locations, with occasional inland records from chalk downs in southern England.

Adult moths fly freely at dusk. The larvae feed on the seeds of the foodplant in July and August.

Similar Species: *Cochylis nana* Page 15

Tortricidae: Cochylinae

Cochylis nana (Haworth, 1811) [0968]

Photo © Neil Sherman

Photo © Jon Clifton

Wingspan: 9-12mm
Flight Period:
May - June
Foodplant:
Downy Birch, Silver Birch
National Status:
Common

The very small size of this species should rule out the superficially similar **Cochylis pallidana**. Also characterised by the diffuse grey-black median fascia appearing confluent with the weak basal fascia (**A**). Note, the strength of the markings can vary with some individuals.

A fairly common and widely distributed species in England and Wales wherever established birch woods are found. Locally common in Scotland and scarce in Ireland.

Adult moths fly actively at dusk. The larvae feed on the catkins of the foodplant from July to October.

Similar Species: Cochylis pallidana Page **14**

Tortricidae: Tortricinae

Acleris variegana ([Denis & Schiffermüller], 1775) [1048]
Garden Rose Tortrix

Photo © Allan Drewitt

Photo © Jon Clifton (form - asperana)

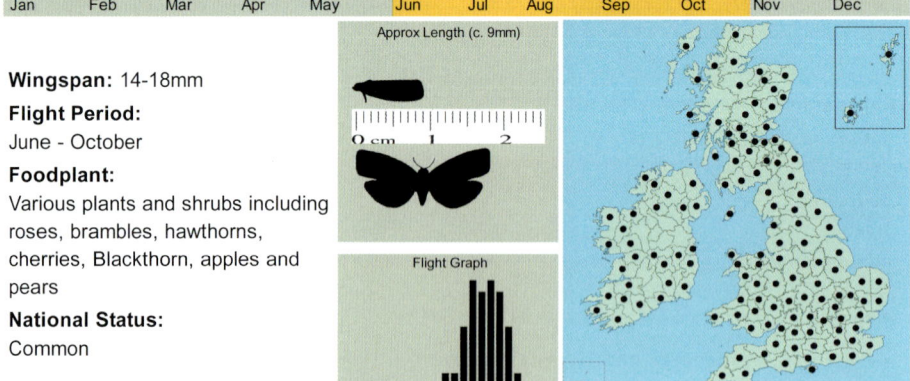

Wingspan: 14-18mm

Flight Period:
June - October

Foodplant:
Various plants and shrubs including roses, brambles, hawthorns, cherries, Blackthorn, apples and pears

National Status:
Common

Although this species in itself is highly variable, mention here is only of the black and white forms to keep within the context of this guide.

Despite there being several black and white forms, neither should cause identification challenges. The forewing being demarcated centrally of which the basal half being white (**A**) and the distal (outer) half being ferruginous/black (**B**). Some forms will show a large dark patch on the dorsum before the middle.

Common and widespread in gardens, hedgerows, woodlands, fens and orchards throughout the British Isles.

Adults are easily disturbed during the day, later coming to light and occasionally to sugar. The larvae feed on the leaves of various low growing plants and shrubs, especially species of *Rosa*, from May to early July.

Tortricidae: Olethreutinae
Celypha woodiana (Barrett, 1882) [1066]

Photo © Mark Parsons

Photo © Mike Wall (Mark Parsons Collection)

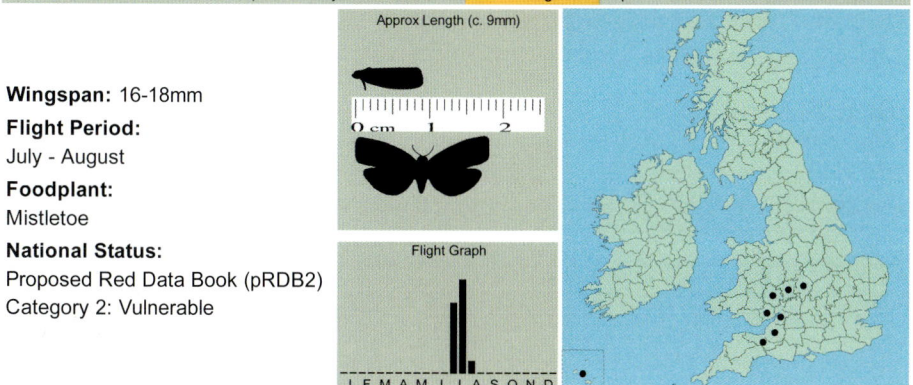

Wingspan: 16-18mm
Flight Period:
July - August
Foodplant:
Mistletoe
National Status:
Proposed Red Data Book (pRDB2) Category 2: Vulnerable

Distinctive. Markings similar to **Apotomis lineana** but is smaller possessing richer forewing markings. The median fascia being well developed costally with dark blue-grey, becoming interupted dorsally with warm olive markings (**A**).

Extremely rare species, virtually restricted to mature apple orchards in south-west England. (one old record from the Jersey, Channel Isles in 1975)

Adult moths are on the wing towards dusk, later coming to light. The larvae mine the leaves of the foodplant from September, forming a large blotch or blister mine.

Priority species under the UK Biodiversity Action Plan.

Similar Species: *Apotomis lineana* Page **26**

Tortricidae: Olethreutinae

Piniphila bifasciana (Haworth, 1811) [1079]

Photo © Allan Drewitt   Photo © Jon Clifton

Wingspan: 12-16mm
Flight Period:
June - July
Foodplant:
Scots Pine, Maritime Pine
National Status:
Local

Almost parallel-sided antemedian and median fascia (A) with a slight indentation on the outer edge (B) (this can be hard to see if worn), diffuse pale ochreous subterminal patch to terminal area of the forewing (C). Can be confused with *Lobesia abscisana* but shows more ochreous tones and is less boldly marked.

Local in coniferous woodlands throughout much of the British Isles. Widely distributed in the southern counties, its range extending to Ireland and northwards to Scotland where it is rare.

Adult moths fly freely at dusk and later come readily to light. The larvae feed in the shoots and blossom of the foodplant in May and June.

Similar Species: *Lobesia abscisana* Page 35

Tortricidae: Olethreutinae

Hedya pruniana (Hübner, 1799) [1082]
Plum Tortrix

Photo © Allan Drewitt

Photo © Jon Clifton

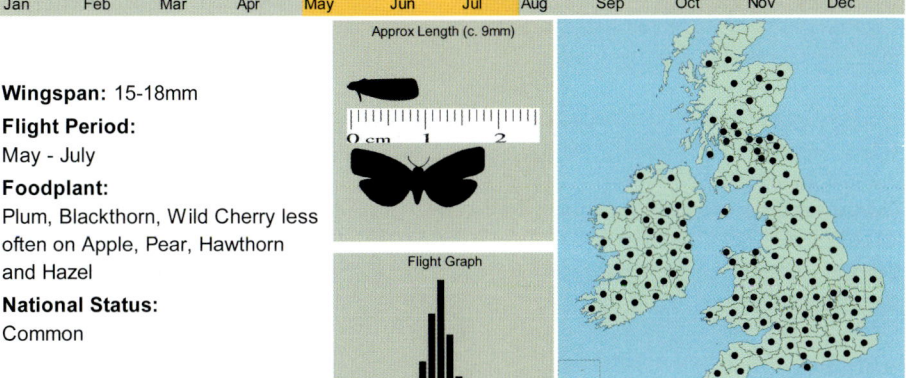

Wingspan: 15-18mm

Flight Period:
May - July

Foodplant:
Plum, Blackthorn, Wild Cherry less often on Apple, Pear, Hawthorn and Hazel

National Status:
Common

Resembles ***Hedya nubiferana*** but is slightly smaller. The two dark postmedian dots are situated away (distad) from the angle of the median fascia (**A**). Some amount of variation occurs in the forewing markings which can become weak in some populations.

Widespread and common in woodlands, hedgerows and gardens throughout the British Isles.

Adult moths fly freely during late evening and later come to light. The larvae feed in the shoots, and later in the leaves of various foodplants in April and May.

Similar Species: *Hedya nubiferana* Page 20   *Hedya ochroleucana* Page 21

Tortricidae: Olethreutinae

Hedya nubiferana (Haworth, 1811) [1083]
Marbled Orchard Tortrix

Photo © Allan Drewitt   Photo © Jon Clifton

Wingspan: 15-21mm

Flight Period:
June - August

Foodplant:
Various trees and shrubs including Hawthorn, Blackthorn and Wild Cherry

National Status:
Common

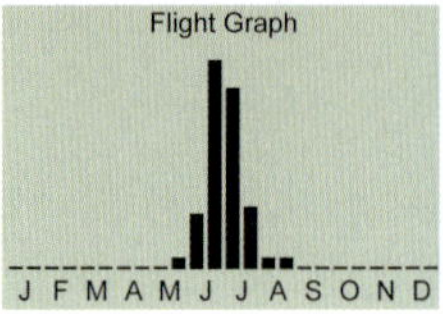

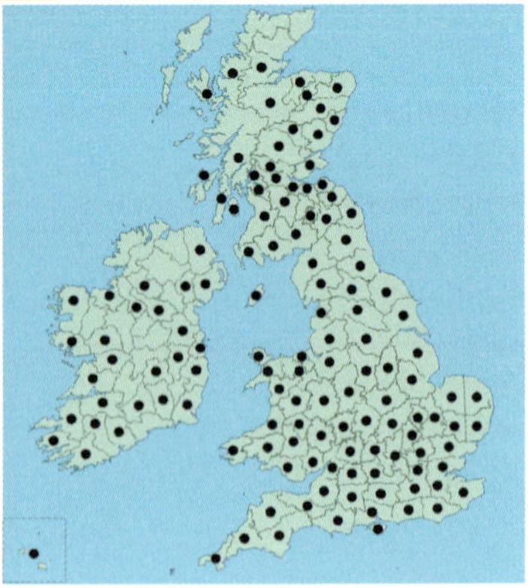

Similar to **Hedya pruniana** but generally larger. The two dark postmedian dots are situated above (costad) the angle of the median fascia (A).

Common and widespread throughout the British Isles in gardens, woodlands and on heathland.

Adult moths fly actively towards dusk and later come to light. The larvae feed in the spun shoots, leaves and flowers of various foodplants from August to May, overwintering in a silken hibernaculum and recommencing feeding in the spring.

Similar Species: *Hedya pruniana* Page **19**   *Hedya ochroleucana* Page **21**

Tortricidae: Olethreutinae

Hedya ochroleucana (Frölich, 1828) [1084]

Photo © Perry Hampson

Photo © Jon Clifton

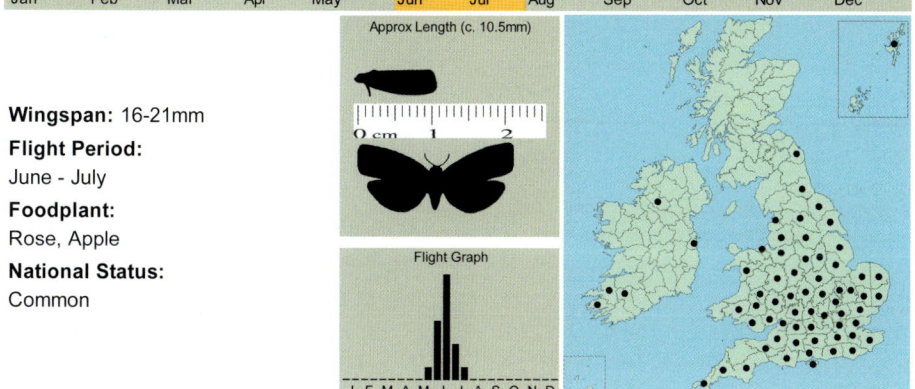

Wingspan: 16-21mm
Flight Period:
June - July
Foodplant:
Rose, Apple
National Status:
Common

The sandy-ochreous cream colouration and a lack of any greyish tones in the outer (distal) part of the forewing is diagnostic and should rule out other similar species, (A).

When photographing this species, depending on lighting and the camera settings, the ochreous tones can easily be washed out leading to confusion.

A fairly common species in gardens, woodlands and hedgerows in southern England and Wales, becoming scarce further north. Local and uncommon in Ireland.

Adult moths may be seen flying at dusk, later coming to light. The larvae feed on the leaves of the foodplant from April to mid June.

Similar Species: *Hedya pruniana* Page 19   *Hedya nubiferana* Page 20

Tortricidae: Olethreutinae

Hedya salicella (Linnaeus, 1758) [1086]

Photo © Allan Drewitt   Photo © Jon Clifton

Wingspan: 19-24mm

Flight Period:
June - September

Foodplant:
White Willow, Goat Willow, Aspen, Black Poplar

National Status:
Local

A large and distinctive species showing a white thorax and tegula (A), and a large unmarked white dorsal area to forewing (B).

Locally widespread in England and Wales becoming scarce further north, recorded in only Ayrshire and Berwickshire in Scotland. Occurs in marshy places, open woodlands and occasionally parks and gardens.

Adult moths fly actively at sunset and later come to light. The larvae living in spun shoots or folded leaves of the foodplant in May and June.

Tortricidae: Olethreutinae

Metendothenia atropunctana (Zetterstedt, 1840) [1085]
(=Hedya dimidiana)

Photo © Neil Sherman

Photo © Jon Clifton

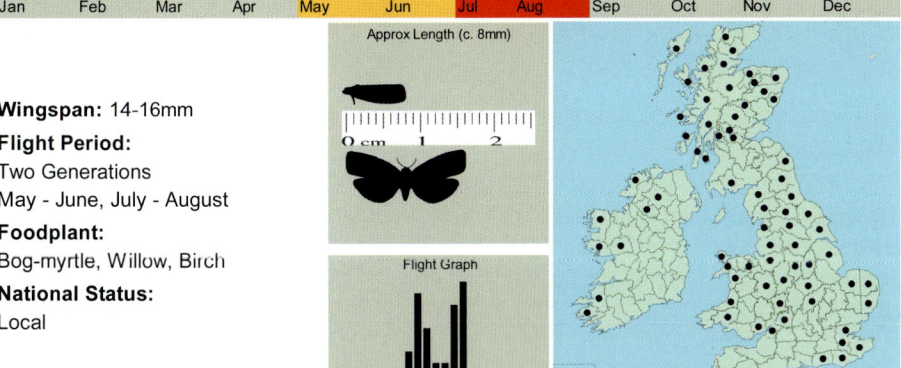

Wingspan: 14-16mm

Flight Period:
Two Generations
May - June, July - August

Foodplant:
Bog-myrtle, Willow, Birch

National Status:
Local

A fairly small species with a slight ochreous ground colour to the forewing, the black median fascia showing a rufous tinge (A) (stronger in some populations), a distinctive black spot distad of the median fascia (B), ochreous markings in subterminal fascia (C).

Occurring around mosses and damp heaths, mainly in the north of England and Scotland, but is also recorded from a few scattered localities in southern England and the west of Ireland.

Adult moths fly during the evening. The larvae feed in the terminal shoots of *Myrica* (Bog-myrtle) and spun leaves of *Salix* and *Betula* from August to October.

Similar Species: *Hedya pruniana* Page 19   *Hedya nubiferana* Page 20

Tortricidae: Olethreutinae

Apotomis semifasciana (Haworth, 1811) [1089]

Photo © Jim Wheeler   Photo © Jon Clifton

Wingspan: 17-20mm
Flight Period:
July - August
Foodplant:
Sallow
National Status:
Local

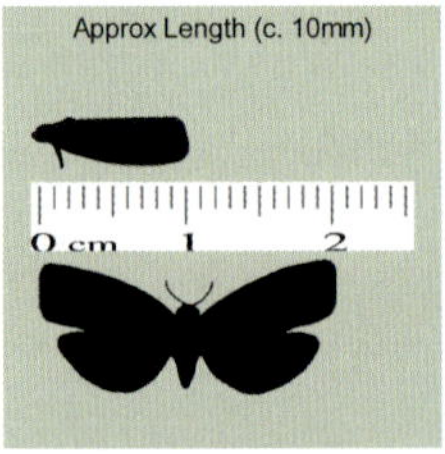

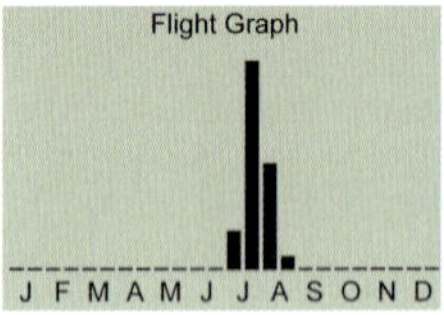

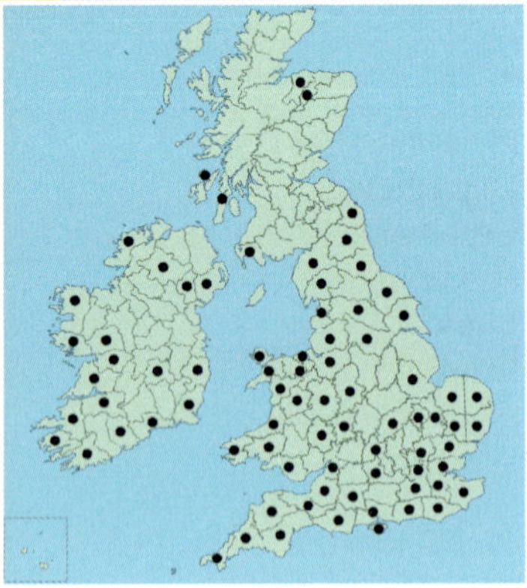

Virtually indistinguishable from **Apotomis infida** but shows a less angulated outer edge to the sub-basal fascia (**A**) and a less outwardly oblique inner edge of the median fascia (**B**) but this is very hard to discern.

Also similar to **Apotomis lineana**, but the dorsal part of forewing slate grey (**C**), lacking the dark greyish/brown suffusion seen in that species.

There can be variation in the ground colour and markings in some specimens so care is required.

Widely distributed in the British Isles but rare in Scotland. Occurring in woodlands, fens, hedgerows and open situations.

Adult moths fly at dusk, later coming to light and occasionally sugar. The larvae feed in the catkins and later in spun leaves of various species of *Salix* in May and June.

Similar Species: *Apotomis lineana* Page 26   *Apotomis infida* Page 25

Tortricidae: Olethreutinae
Apotomis infida (Heinrich, 1926) [1090]

Photo © James Roberts

Photo © SIP Haapala 2007

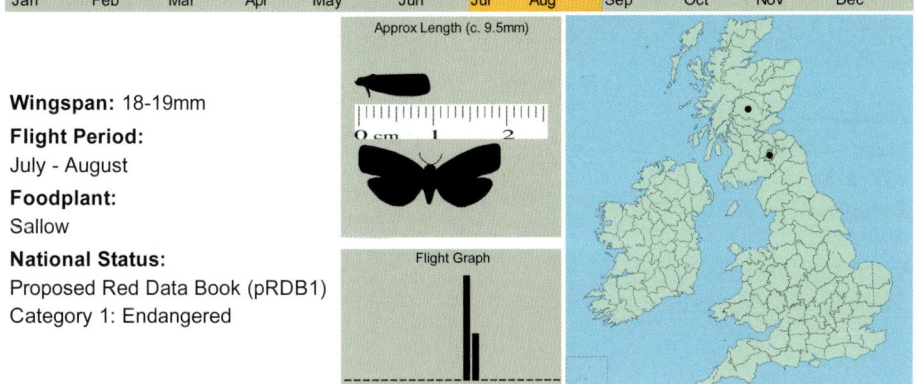

Wingspan: 18-19mm
Flight Period:
July - August
Foodplant:
Sallow
National Status:
Proposed Red Data Book (pRDB1) Category 1: Endangered

As the scientific name *infida - infidus* implies, "not to be trusted", a very variable species and can be deceptive. The ground colour in some specimens can be lighter or darker than others.

Very similar to **Apotomis semifasciana**. Generally tends to show a more angulate outer edge to the sub-basal fascia (A) and more outwardly oblique inner edge of the median fascia (B) but this is very slight and any potential records would have to be supported by a voucher specimen.

This species has been recorded at only two known sites in Scotland. Two specimens taken at Loch Rannoch in the collection of F.G. Whittle, dated the 11th and 14th July, 1919 (Whittle, 1920: 11), were originally recorded as *A. semifasciana*, only later to be identified as *A. infida*. Further specimens were taken at Ettrick, Selkirk in early August 1979 and mid-July 1980 by K. Bland and E.C. Pelham-Clinton. (pers. comm. K. Bland).

Further searches should be considered as little is known of the biology of this species. Old, well established stands of sallow-carr seems to be a preferred habitat.

Similar Species: *Apotomis semifasciana* Page 24

Tortricidae: Olethreutinae

Apotomis lineana ([Denis & Schiffermüller], 1775) [1091]

Photo © Allan Drewitt

Photo © Mike Wall (Brian Elliott Collection)

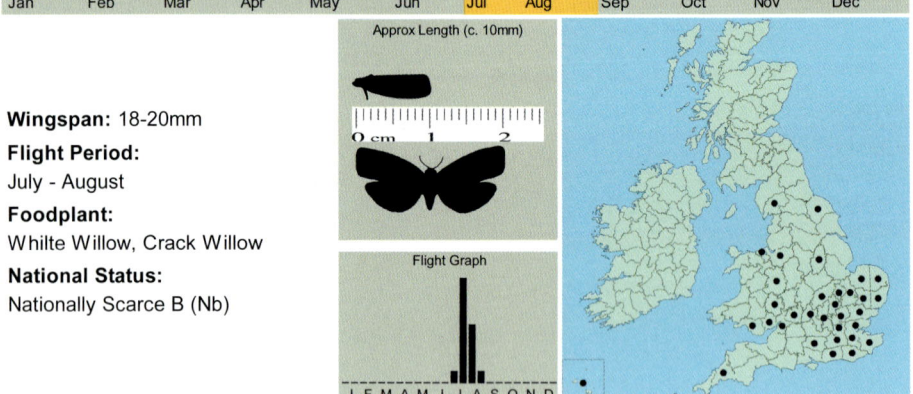

Wingspan: 18-20mm
Flight Period:
July - August
Foodplant:
Whilte Willow, Crack Willow
National Status:
Nationally Scarce B (Nb)

Resembles lighter forms of **A. semifasciana**, but differs in showing a white/greyish-white ground colour to the forewing and a dark grey/brown suffusion along the dorsum (A).

Found in wet habitats including river banks, bogs and water meadows. Uncommonly recorded in southern England and Wales with a few scattered records from the north of England.

Adult moths are seldom seen in flight, occasionally found at rest on tree trunks. The larvae feed between the spun leaves of the foodplant in May.

Similar Species: *Apotomis semifasciana* Page 24   *Apotomis turbidana* Page 27

Tortricidae: Olethreutinae

Apotomis turbidana (Hübner, 1825) [1092]

Photo © Allan Drewitt

Photo © Jon Clifton

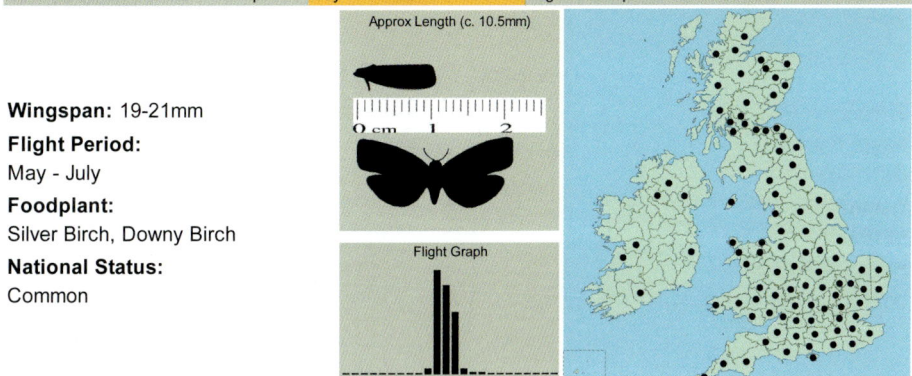

Wingspan: 19-21mm

Flight Period:
May - July

Foodplant:
Silver Birch, Downy Birch

National Status:
Common

A thick white streak to basal area of forewing, mottled with grey and black scales (**A**) extending onto the costa (**B**) which is diagnostic. Paler apical area and straighter outer edge to the lower part of median fascia seperating it from **Apotomis sororculana**. Some minor variation can be found in the intensity of dark markings, but these should cause little confusion.

Occurring in woodlands and heathland where birch trees and bushes may be found, a relatively common species throughout the British Isles.

Adult moths fly from dusk and later come to light. The larvae feed in the spun leaves of the foodplant in May.

Similar Species: *Apotomis lineana* Page 26   *Apotomis sororculana* Page 30

Tortricidae: Olethreutinae

Apotomis betuletana (Haworth, 1811) [1093]

Photo © Allan Drewitt   Photo © Jon Clifton

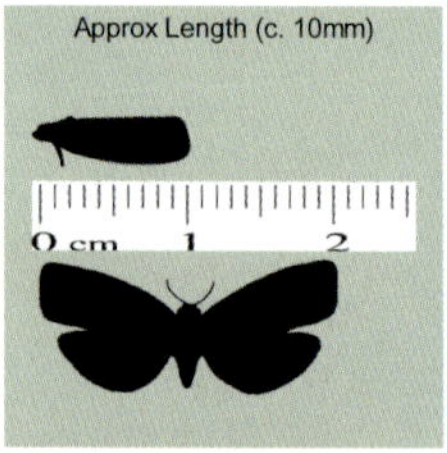

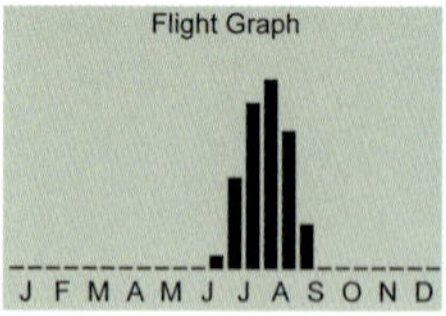

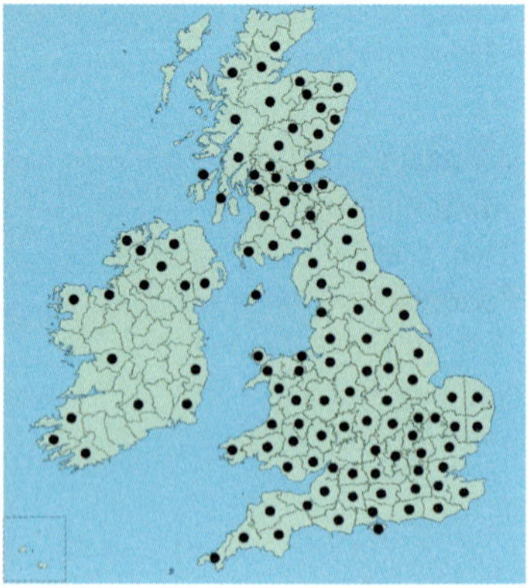

Wingspan: 16-20mm
Flight Period:
June - September
Foodplant:
Birch
National Status:
Common

Similar to *Apotomis capreana* but has a straighter median fascia (A), and lacks the black dots in the subterminal fascia (B). Could also resemble *Apotomis sororculana* but that species shows a more angulated median fascia.

Common and widespread amongst birch throughout the British Isles.

Adult moths fly from dusk, later coming to light and also sugar. The larvae feed between the spun leaves of the foodplant in May and June.

Similar Species: *Apotomis turbidana* Page 27   *Apotomis capreana* Page 29

Tortricidae: Olethreutinae

Apotomis capreana (Hübner, 1817) [1094]

Photo © Allan Drewitt   Photo © Jon Clifton

| Jan | Feb | Mar | Apr | May | Jun | Jul | Aug | Sep | Oct | Nov | Dec |

Wingspan: 17-22mm
Flight Period:
June - October
Foodplant:
Goat Willow
National Status:
Local

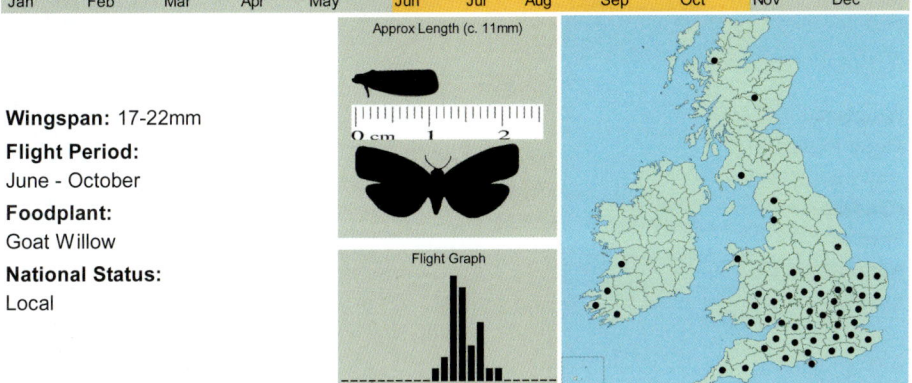

Similar to ***Apotomis betuletana*** but shows a stronger hook-like indentation to the outer edge of the median fascia (**A**), this can be hard to interpret as *A. betuletana* shows a weak indentation so care should be exercised. The median fascia is more concave (not straight as in *A. betuletana*) appearing to curve along the costal edge of the forewing (**B**), stronger black dots set in the subterminal fascia (**C**).

Apotomis sauciana is smaller showing a deeper blue-grey ground colour and a dark brown-black subterminal fascia extending into the apical area.

Locally recorded in open woodlands, marshes and fens. Occurring mainly in the southern counties of England and Wales. A few scattered records from Scotland and south-west Ireland.

The adult flies from dusk and comes to light. The larvae feed in the spun leaves of the foodplant in April and May.

Similar Species: *Apotomis betuletana* Page **28**   *Apotomis sauciana* Page **31**

Tortricidae: Olethreutinae

Apotomis sororculana (Zetterstedt, 1839) [1095]

Photo © Jim Wheeler   Photo © Mike Wall (John Langmaid Collection)

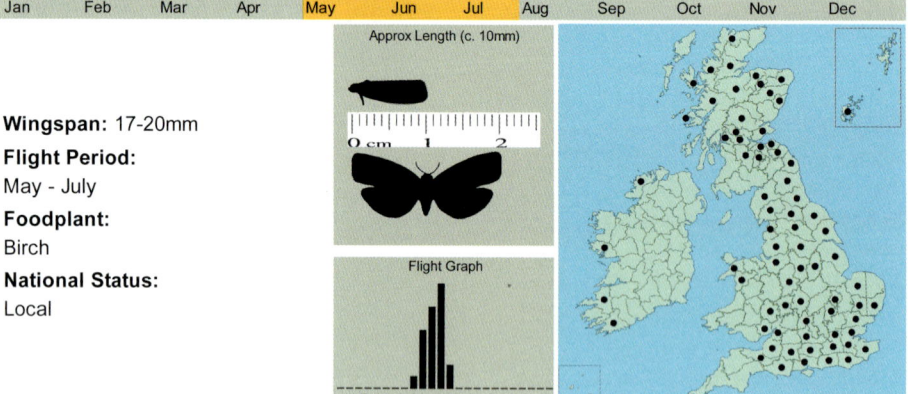

Wingspan: 17-20mm
Flight Period:
May - July
Foodplant:
Birch
National Status:
Local

Can resemble other species of *Apotomis*, especially **Apotomis turbidana** and **A. betuletana** but recognised by the more angulated median fascia which is less indented on its outer edge (**A**) and by the black apex and cilia (**B**). Hindwing lighter (**C**).

This species has a wide distribution throughout the British Isles, but is generally commoner in the north and rare in Wales and Ireland.

Adult moths fly towards dusk, later coming to light. The larvae feed between the leaves of the foodplant from July to September.

Similar Species: *Apotomis turbidana* Page **27**   *Apotomis betuletana* Page **28**

Tortricidae: Olethreutinae
Apotomis sauciana (Frölich, 1828) [1096]

Photo © Jim Wheeler   Photo © Jon Clifton

Wingspan: 13-16mm
Flight Period:
June - August
Foodplant:
Bilberry, Bearberry
National Status:
Common

Similar to **Apotomis capreana** but distinguished by its smaller size, deeper blue-grey ground colour of the forewing and the dark brown-black subterminal fascia extending into the apical area (A).

The northern subspecies **A. sauciana grevillana** (red map dots) occurring in Scotland differs from the nominate subspecies by the slightly paler blue-grey ground colour to the basal area of the forewing and the dark markings in the sub-terminal fascia appearing obscure.

Widespread but scattered records mainly from the south and south-west of England, Wales and Scotland.

Adult moths can be found resting openly on the leaves of the foodplant during the day, flying freely in the afternoon and evening when the weather is warm. The larvae feed in the terminal leaves of Bilberry form April to June. In Scotland larvae also feed on Bearberry.

Similar Species: *Endothenia gentianaeana* Page 32   *Apotomis capreana* Page 29

Tortricidae: Olethreutinae

Endothenia gentianaeana (Hübner, 1799) [1097]

Photo © Mike Crewe   Photo © Jon Clifton

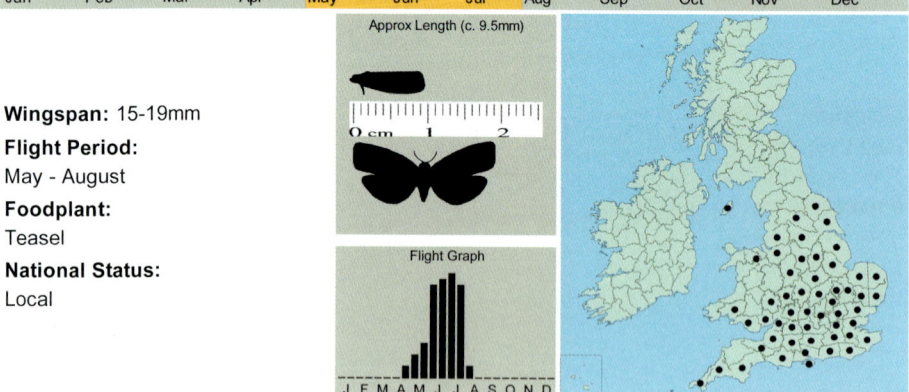

Wingspan: 15-19mm
Flight Period:
May - August
Foodplant:
Teasel
National Status:
Local

On average much larger than the other two UK species of *Endothenia* but some small specimens do occur and can cause problems. *E. gentianaeana* usually shows more violet or greyish areas to the tornal and basal areas of the forewing (**A**) but these markings can vary considerably, especially the extent of white in the basal area and the black spots present. An inconspicuous white discocellular spot present on the outer margin of the median fascia (**B**) (this can be hard to discern depending on the amount of markings around it) subterminal fascia weakly violet with several scattered small black dots (**C**).

Occurring on rough uncultivated ground, hedgerows and embankments where common teasel grows. Widely distributed in southern England becoming scarce further north.

Adult moths are seldom seen on the wing. The larvae feed in the seed-heads of the foodplant in September and October, overwintering until the following spring. Moths can be reared by collecting the teasel heads in the spring.

Similar Species: *Endothenia oblongana* Page **33**   *Endothenia marginana* Page **34**

Tortricidae: Olethreutinae

Endothenia oblongana (Haworth, 1811) [1098]

Photo © Jim Wheeler   Photo © Jon Clifton

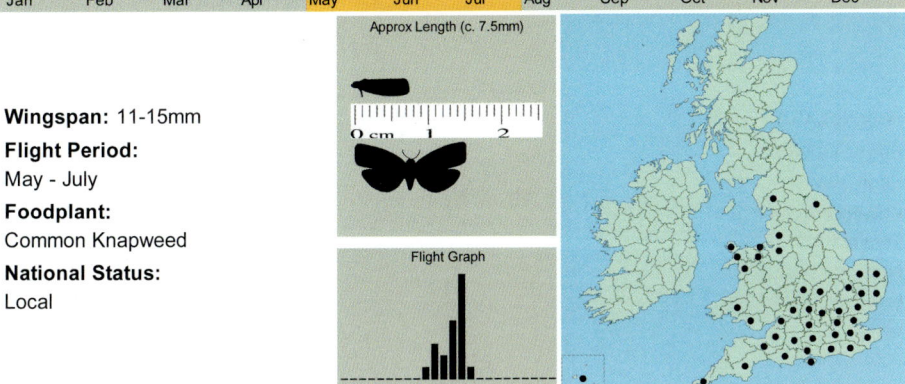

Wingspan: 11-15mm
Flight Period:
May - July
Foodplant:
Common Knapweed
National Status:
Local

Resembles a medium-sized **Endothenia gentianaeana**. Broadly suffused apical area reducing the whitish band distad of the median fascia (A), a stronger dilated subterminal fascia lacking any violet tones (B). Some specimens may need to be retained for determination.

Occurring in well-drained uncultivated ground, such as waysides, embankments, chalk downs and quarries. Locally recorded mainly in the southern and south-eastern counties of England and Wales.

Adult moths fly freely during the evening and come to light. The larvae is found in the roots of the foodplant from September to May.

Similar Species: Endothenia gentianaeana Page **32**   Endothenia marginana Page **34**

Tortricidae: Olethreutinae

Endothenia marginana (Haworth, 1811) [1099]

Photo © Tristan Bantock   Photo © Jon Clifton

| Jan | Feb | Mar | Apr | May | Jun | Jul | Aug | Sep | Oct | Nov | Dec |

Wingspan: 11-16mm

Flight Period:
May - August

Foodplant:
Betony, hemp-nettles, louseworts, Yellow Rattle, Ribwort Plantain, Teasel

National Status:
Local

Males can be easily distinguished from both sexes of **Endothenia gentianaeana** and **E. oblongana** by its white hindwings (**A**). Both sexes can be separated from *E. gentianaeana* by the narrower area of whitish ground colour distad of the median fascia (**B**). Some specimens, especially females, may need retaining for identification.

Occurring in rough meadows and grassland, waysides, embankments, damp woods, boggy heaths and fens. Widely distributed and locally common, occurring throughout the British Isles.

Adult moths fly low down in suitable habitat during the evening and come to light. The larvae feed in the flower and seed-heads of various foodplants from September to June, overwintering in the seed capsule.

Similar Species: *Endothenia gentianaeana* Page 32   *Endothenia oblongana* Page 33

Tortricidae: Olethreutinae

Lobesia abscisana (Doubleday, 1849) [1108]

Photo © Allan Drewitt

Photo © Jon Clifton

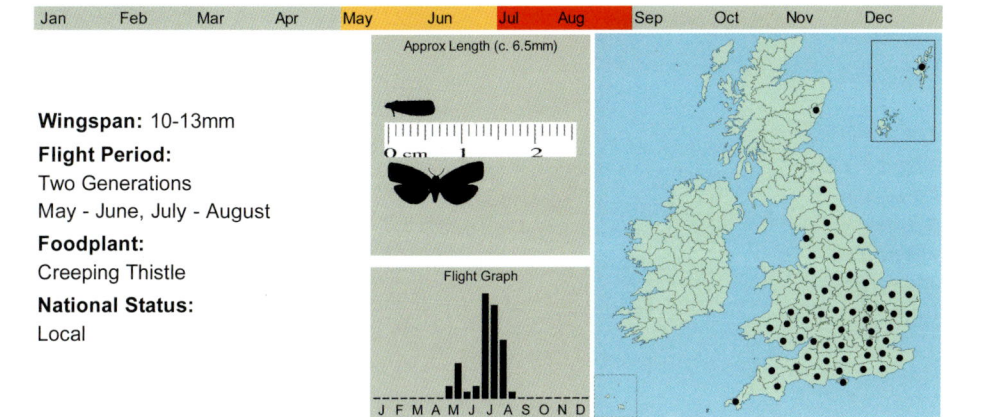

Wingspan: 10-13mm

Flight Period:
Two Generations
May - June, July - August

Foodplant:
Creeping Thistle

National Status:
Local

Could be confused with **Piniphila bifasciana** but has much bolder dark brown markings lacking any ochreous tones. A well defined dark brown basal patch (**A**), median fascia narrow at the costa dilating (expanding) from the middle (**B**), subterminal fascia pale brown (**C**).

Locally widespread and fairly common in the south and south-east of England. Occurring in fields, rough meadows, waysides, flower meadows and waste places. Rare in Scotland with records from Kincardineshire and Shetland.

Adult moths sometimes fly in the afternoon, but mostly from dusk, later coming to light. The larvae feed within the stems of the foodplant in June, with a second generation in August.

Similar Species: *Piniphila bifasciana* Page **18**

Tortricidae: Olethreutinae

Ancylis laetana (Fabricius, 1775) [1123]

Photo © Neil Sherman

Photo © Jon Clifton

| Jan | Feb | Mar | Apr | May | Jun | Jul | Aug | Sep | Oct | Nov | Dec |

Wingspan: 13-18mm
Flight Period:
May - June
Foodplant:
Aspen, Black Poplar
National Status:
Local

Approx Length (c. 9mm)

Flight Graph
J F M A M J J A S O N D

A distinctive species showing a broad white leading edge to forewing (**A**), dark brown-black dorsal area (**B**) and rufous apical area (**C**).

Widely distributed and locally common in open woodlands, parks and gardens in the southern counties of England, becoming scarce further north and west. Very local in Scotland and apparently unknown from Ireland.

Adult moths may be seen flying around Aspen and Black Poplar at sunset, later coming to light. The larvae spin the leaves of the foodplant from July to September, overwintering until the following spring.

Tortricidae: Olethreutinae

Epinotia subocellana (Donovan, 1806) [1132]

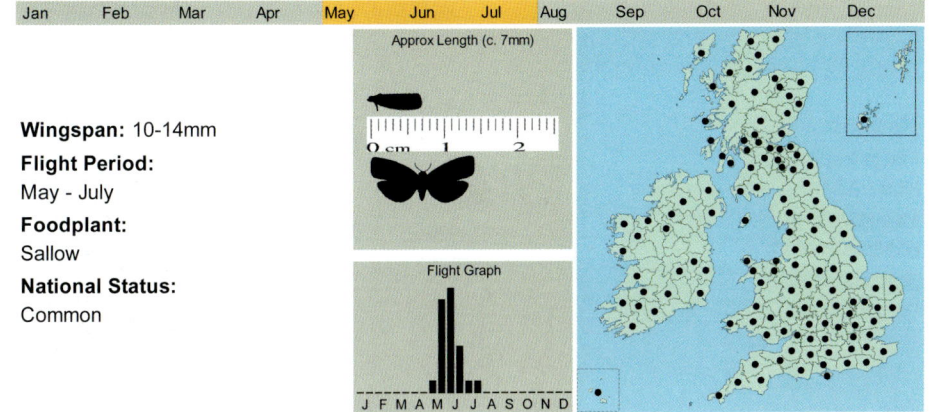

Photo © Rob Lee

Photo © Jon Clifton

| Jan | Feb | Mar | Apr | May | Jun | Jul | Aug | Sep | Oct | Nov | Dec |

Wingspan: 10-14mm
Flight Period:
May - July
Foodplant:
Sallow
National Status:
Common

The basal area of forewing (**A**) diffuse and weakly speckled with brown and blackish markings, median fascia, pre-tornal area and ocellus mixed with rufous, black and metallic blue (**B**), subterminal fascia and apical spot rufous brown and weakly defined (**C**) (more obvious in fresh specimens).

Occurring in woodlands, banks, wet heaths, fens, bogs and other situations where sallows are plentiful. Common and widely distributed throughout the British Isles.

Adult moths fly actively around sallows from dusk and later come to light. The larvae feed in the spun leaves of the foodplant from August to October.

Tortricidae: Olethreutinae

Epinotia bilunana (Haworth, 1811) [1133]

Photo © Allan Drewitt

Photo © Jon Clifton

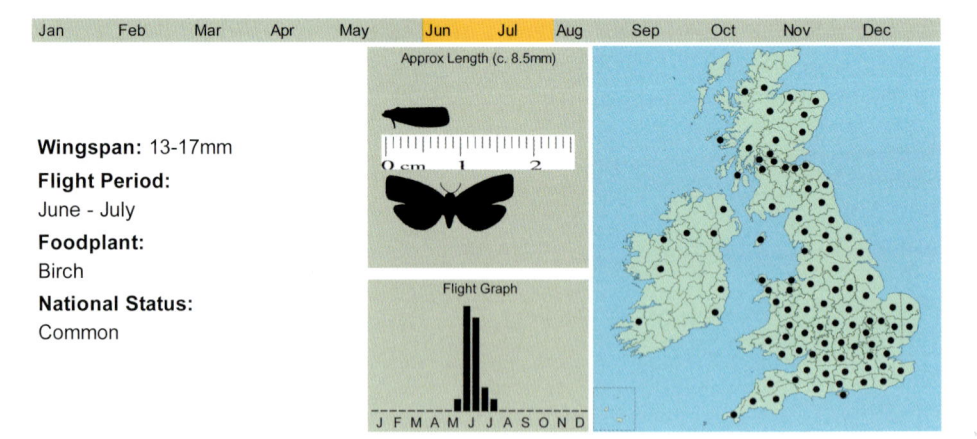

| Jan | Feb | Mar | Apr | May | Jun | Jul | Aug | Sep | Oct | Nov | Dec |

Wingspan: 13-17mm
Flight Period:
June - July
Foodplant:
Birch
National Status:
Common

Separated from *Epinotia ramella* by the triangular patch on the dorsum (**A**) which is much weaker and diffuse in markings, appearing faded (beware worn examples of *E. ramella*). Overall, it is a paler, whiter moth.

Minor variation occurs in the strength of the blackish markings.

Widely distributed and generally common throughout the British Isles.

The adult moth can often be found resting on the trunks of birch trees during the day, later flying from dusk and coming to light. The larvae feed in the catkins of the foodplant from September to April.

Flies earlier in the season than the similar *E. ramella*

Similar Species: *Epinotia ramella* Page **39**

Tortricidae: Olethreutinae

Epinotia ramella (Linnaeus, 1758) [1134]

Photo © Allan Drewitt

Photo © Jon Clifton

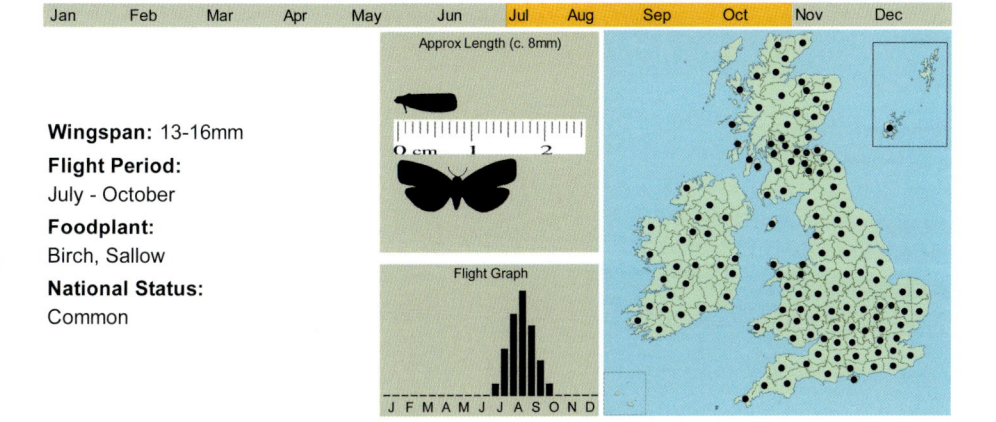

Wingspan: 13-16mm
Flight Period:
July - October
Foodplant:
Birch, Sallow
National Status:
Common

Can be confused with **Epinotia bilunana** but shows a solid dark triangular patch on the dorsum (**A**) (beware worn specimens). Overall, it is a darker, browner moth.

The form *costana* occurs in which the forewing is almost entirely blackish brown other than the area along the costa.

Common in birch woodland throughout much of the British Isles.

Adult moths fly from dusk around the foodplant and later come to light. The larvae feed within the twigs and catkins of the foodplant in April and May.

Flies later in the year than the similar *E. bilunana*.

Similar Species: *Epinotia bilunana* Page 38

Tortricidae: Olethreutinae
Epinotia demarniana (Fischer von Röslerstamm, 1840) [1135]

Photo © Allan Drewitt

Photo © Jon Clifton

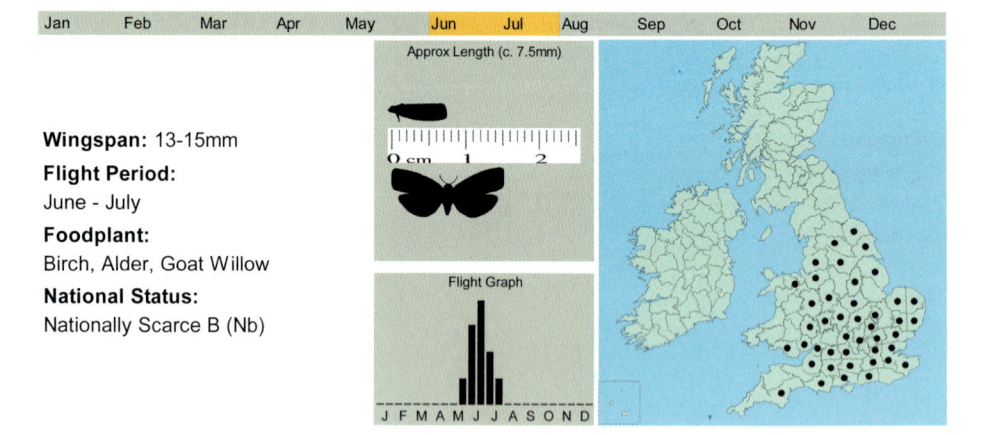

| Jan | Feb | Mar | Apr | May | Jun | Jul | Aug | Sep | Oct | Nov | Dec |

Approx Length (c. 7.5mm)

Wingspan: 13-15mm
Flight Period:
June - July
Foodplant:
Birch, Alder, Goat Willow
National Status:
Nationally Scarce B (Nb)

Flight Graph

Could be confused with similar *Epiblema* but is usually smaller and narrower winged. Head and palps ochreous-white (**A**), dark border running parallel along costa (**B**), median fascia oblique (not straight) (**C**), a large white blotch between basal patch and median fascia, and above the tornus (**D**).

A locally distributed and rather scarce species in the British Isles, recorded mainly in southern and eastern England. Occurring in woodlands, heathland, fens and bogs.

Adult moths fly high around the foliage in the evening, coming to light after dusk. The larvae feed within the catkins of the foodplant from September to May.

Similar Species: *Epiblema scutulana* Page 53

Tortricidae: Olethreutinae

Epinotia trigonella (Linnaeus, 1758) [1151]
(=*stroemiana*)

Photo © Mike Wall

Photo © Jon Clifton

Wingspan: 16-21mm

Flight Period:
August - September

Foodplant:
Birch

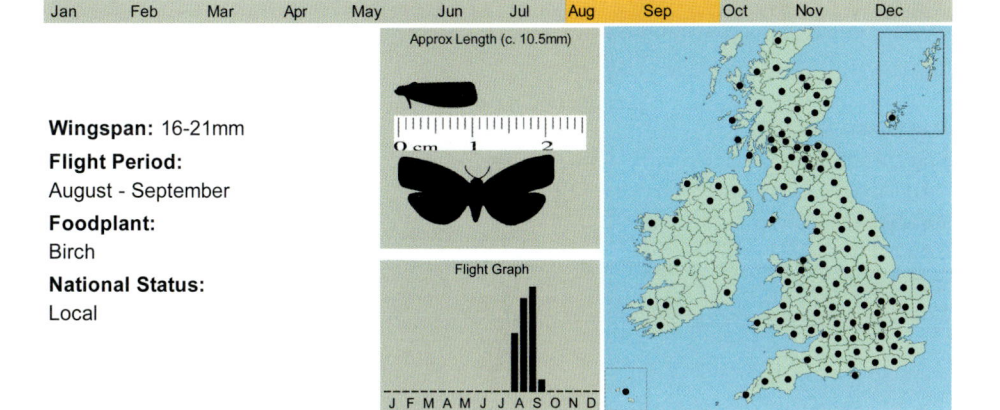

National Status:
Local

Superficially resembles *Epinotia demarniana* but that species is smaller and flies earlier in the season. Dark border running parallel along costa (A), two white blotches along dorsum and in the tornal area (B).

Locally common throughout the British Isles on heathland, mosses, bogs and in open woodlands.

Adult moths fly from dusk and come to light. The larvae feed in spun leaves of the foodplant in June and July.

Similar Species: *Epinotia demarniana* Page **40**

Tortricidae: Olethreutinae

Gypsonoma aceriana (Duponchel, 1843) [1167]

Photo © Neil Sherman

Photo © Mike Wall (John Langmaid Collection)

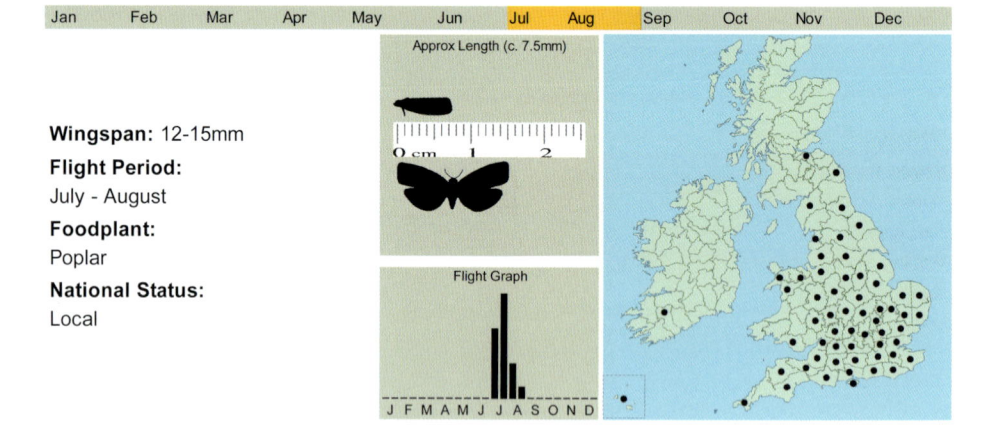

| Jan | Feb | Mar | Apr | May | Jun | Jul | Aug | Sep | Oct | Nov | Dec |

Wingspan: 12-15mm
Flight Period:
July - August
Foodplant:
Poplar
National Status:
Local

Could be confused with *Gypsonoma dealbana* and *Spilonota ocellana*.

Separated from *G. dealbana* by showing a pink/ochreous suffusion to the white ground colour of the forewing (this can be lost with wear), also told by the basal patch being straighter (A).

S. ocellana shows a larger more distinctive triangular black pre-tornal mark and has several distinctive black dashes in the apical area of the ocellus.

Locally common in the south of England becoming scarce further north. The only known Scottish record is of a specimen from Midlothian in 1975. Rare in Ireland. Occurring in open woodlands, hedgerows, parks and gardens where poplars grow.

Adult moths fly from sunset and occasionally come to light. The larvae feed in the shoots of the foodplant in May and June.

Similar Species: *Spilonota ocellana* Page **60**   *Gypsonoma dealbana* Page **44**

Tortricidae: Olethreutinae

Gypsonoma sociana (Haworth, 1811) [1168]

Photo © Jim Wheeler

Photo © Jon Clifton

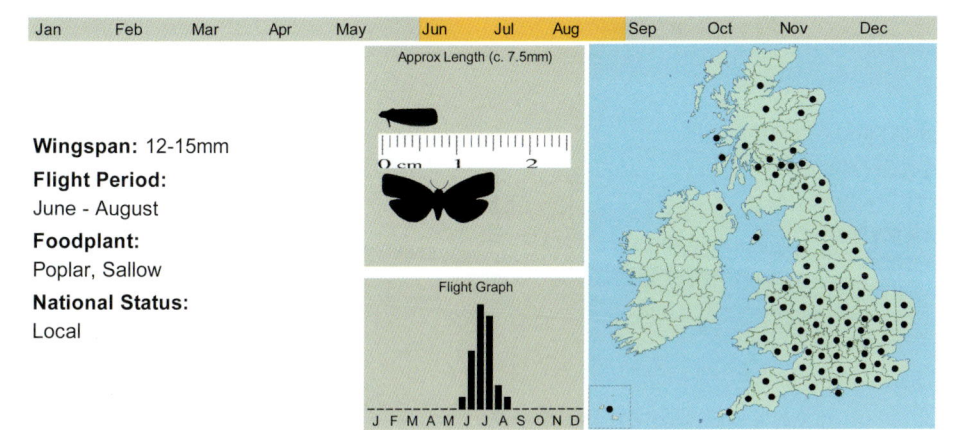

| Jan | Feb | Mar | Apr | May | Jun | Jul | Aug | Sep | Oct | Nov | Dec |

Approx Length (c. 7.5mm)

Wingspan: 12-15mm

Flight Period:
June - August

Foodplant:
Poplar, Sallow

National Status:
Local

Flight Graph

J F M A M J J A S O N D

Distinguished with care from **Gypsonoma dealbana** by the **clear white** frons (**A**) (see inset), further separated by the overall colouration of the forewings being cleaner and more 'pied'.

Locally fairly common over much of the British Isles in open woodlands, gardens, rivers and streams. Rare in Ireland.

Adult moths fly actively around the trees during the evening and later come to light. The larvae feed in the shoots of the foodplant from September to May. After overwintering the larvae feed within a leaf bud until fully grown. On *Salix*, the larvae feed in the catkins after overwintering.

Similar Species: *Gypsonoma dealbana* Page 44

Tortricidae: Olethreutinae

Gypsonoma dealbana (Frölich, 1828) [1169]

Photo © Jim Wheeler

Photo © Jon Clifton

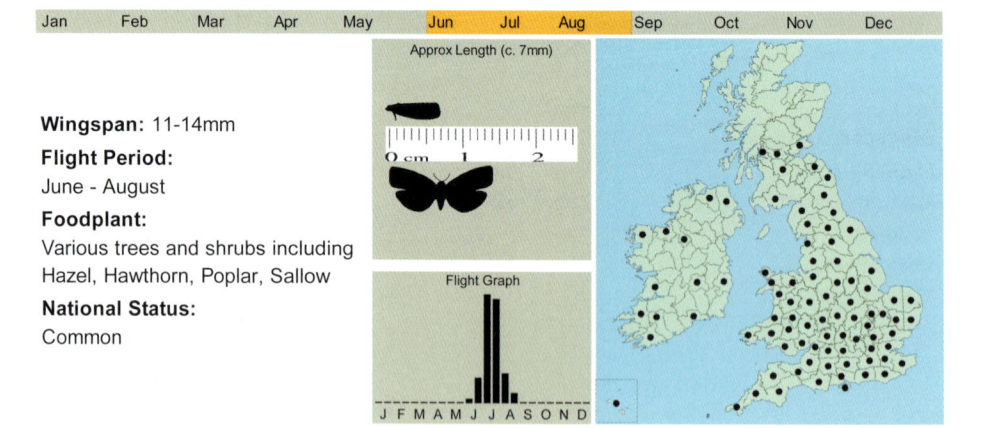

| Jan | Feb | Mar | Apr | May | Jun | Jul | Aug | Sep | Oct | Nov | Dec |

Wingspan: 11-14mm
Flight Period:
June - August
Foodplant:
Various trees and shrubs including Hazel, Hawthorn, Poplar, Sallow
National Status:
Common

Approx Length (c. 7mm)

Flight Graph

J F M A M J J A S O N D

A highly variable species in overall colour and markings which can differ greatly from a white ground colour to brownish/grey making the markings obscure. Can be distinguished from *Gypsonoma sociana* with care by the cream-white/ochreous-white/greyish-white frons (**A**). Interpreting the quoted colour of the frons of *G. dealbana* from the "clear white frons" of *G. sociana* can be problematical and the observer may become easily confused, care is required.
Other critical features consist of a well defined basal patch angled below middle (**B**), median fascia weak showing a black discocellular dash (**C**), rufous apical spot (**D**).

Separated from *Gypsonoma aceriana* by the lack of a pink/ochreous suffusion to the forewing and by the angled basal patch (straighter in *G. aceriana*).

Widespread and common in open woodlands throughout much of the British Isles.

At dusk the adult moths fly high up around the foodplant and later come to light. The larvae feed within buds and young shoots of various foodplants from September to early June, overwintering and eating into the buds, shoots and catkins in the spring.

Similar Species: *Gypsonoma aceriana* Page **42**   *Gypsonoma sociana* Page **43**

Tortricidae: Olethreutinae

Gypsonoma oppressana (Treitschke, 1835) [1170]

Photo © Patrick Clement

Photo © Jon Clifton

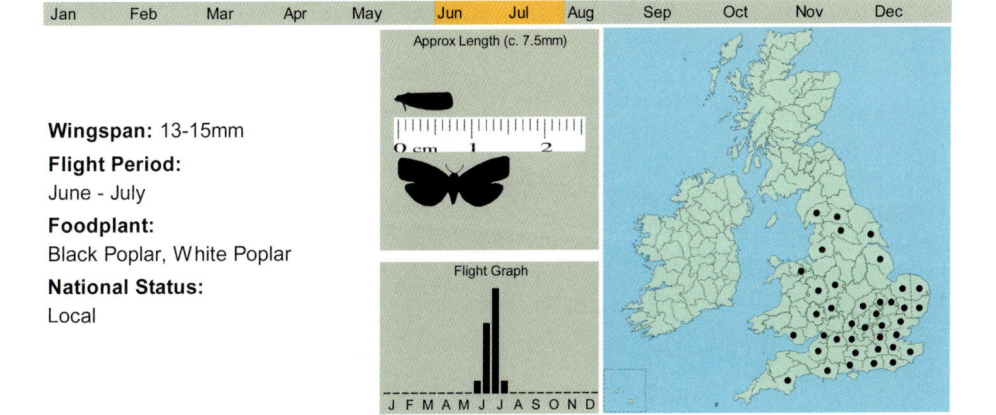

Wingspan: 13-15mm
Flight Period:
June - July
Foodplant:
Black Poplar, White Poplar
National Status:
Local

Ground colour white mixed with grey/brown. The basal patch having the outer edge angulated (**A**), median fascia poorly defined showing grey/brown markings (**B**), a weak apical spot in the subterminal fascia (**C**). There can be slight variation of the coloration and strength of the markings so care is needed, if in doubt, retain the specimen.

Locally recorded throughout much of southern England becoming scattered further north. Found at the edges of woodlands, meadows, gardens and similar situations where poplars grow.

Adult moths fly from sunset and come to light. The larvae feed on the leaves and buds of the foodplant from September to May.

Similar Species: *Gypsonoma dealbana* Page 44

Tortricidae: Olethreutinae

Gibberifera simplana (Fischer von Röslerstamm, 1836) [1173]

R

Photo © Peter Buchner

Photo © Harry Beaumont

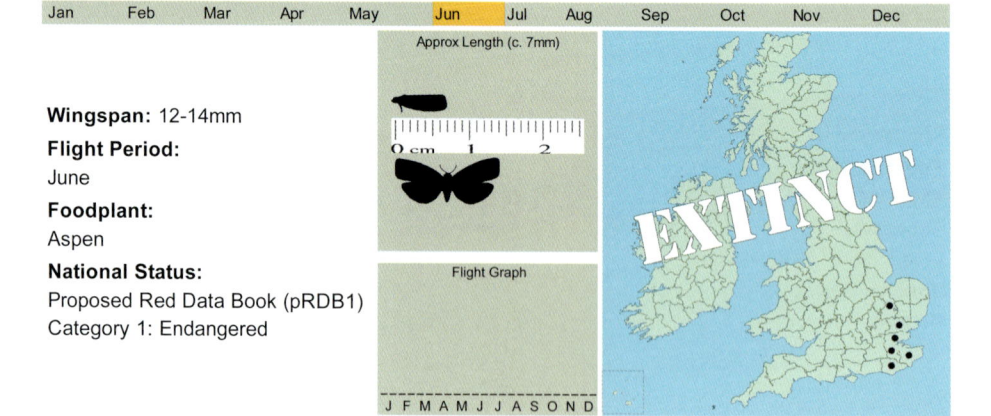

| Jan | Feb | Mar | Apr | May | Jun | Jul | Aug | Sep | Oct | Nov | Dec |

Approx Length (c. 7mm)

Wingspan: 12-14mm

Flight Period:
June

Foodplant:
Aspen

National Status:
Proposed Red Data Book (pRDB1) Category 1: Endangered

Flight Graph

J F M A M J J A S O N D

EXTINCT

Fairly distinctive. Ground colour of forewing ochreous buff. Outer edge of basal patch sharply defined and angular (**A**), median fascia represented by a dark blotch on the costa (**B**), pre-tornal mark weak (**C**) even in fresh specimens. A small but well defined black apical spot (**D**).

Possibly confused with *Gypsonoma aceriana* or *Spilonota ocellana* but flies earlier. Also shows a richer ground colour to the forewing and by the shape of the basal patch and the dark blotch along the costa of the forewing.

Known only from south-east England where it was last recorded in 1930. Former localities include Wicken Fen (Cambs), Colchester and Hadleigh (Essex), Barham, Darenth Wood, Elham and Folkestone (Kent), Abbot's Wood, Brighton, Polegate and Tilgate Forest (Sussex). (Baldwin, 1878; Bradley, 1979; Ferguson, 2003; Agassiz, 2004)

Found in coppiced woodlands, with a preference for new growth and young aspens about 2m in height. This species is now believed to be extinct in the British Isles but further searches should be considered in likely habitat.

Similar Species: *Gypsonoma aceriana* Page **42**   *Spilonota ocellana* Page **60**

Tortricidae: Olethreutinae

Epiblema cynosbatella (Linnaeus, 1758) [1174]

Photo © Allan Drewitt   Photo © Jon Clifton

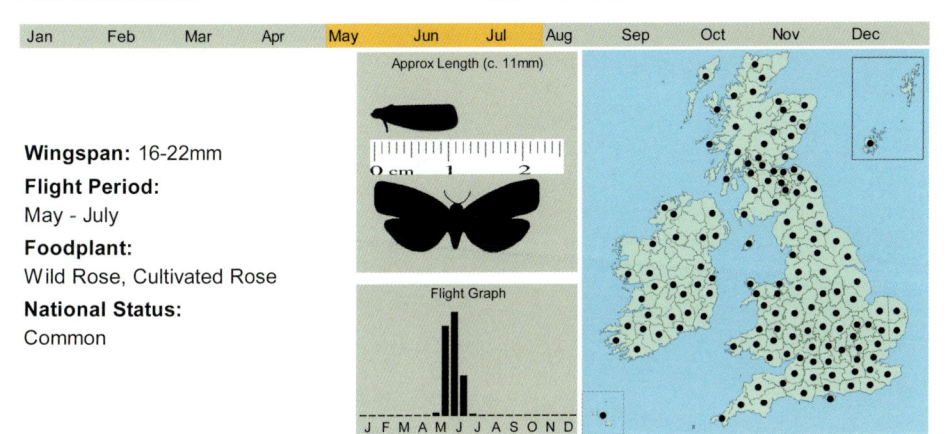

| Jan | Feb | Mar | Apr | May | Jun | Jul | Aug | Sep | Oct | Nov | Dec |

Wingspan: 16-22mm
Flight Period:
May - July
Foodplant:
Wild Rose, Cultivated Rose
National Status:
Common

One of the classic bird dropping tortrix moths and instantly distinguished from other congeners by the yellowish labial palps (**A**) (beware worn examples), basal patch extending along costa (**B**) and ocellus which shows three or four black dots (**C**).

Widespread and common in hedgerows, orchards and gardens throughout the British Isles.

Adult moths fly during late afternoon and evening, seldom moving far from the foodplant, occasionally coming to light. The larvae live within a spun or rolled leaf of the foodplant, feeding in April and May.

Similar Species: *Epiblema trimaculana* Page **48**   *Epiblema rosaecolana* Page **49**

Tortricidae: Olethreutinae

Epiblema trimaculana (Haworth, 1811) [1176]

Photo © Jim Wheeler

Photo © Jon Clifton

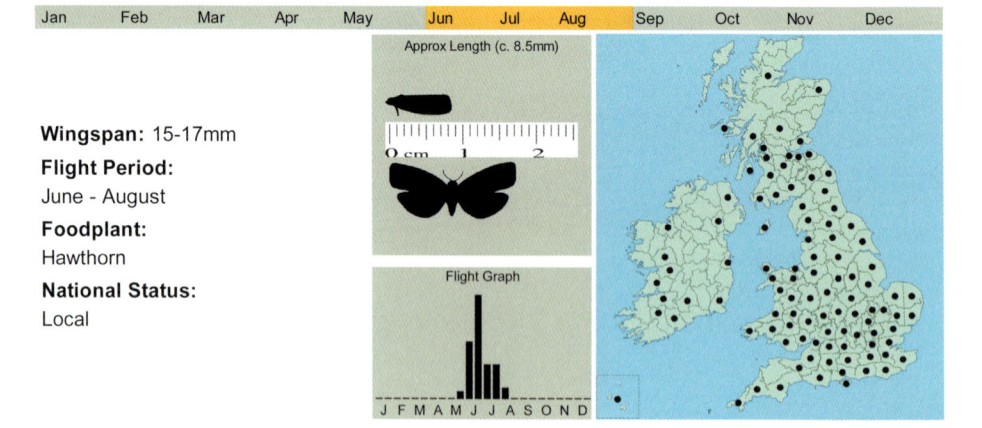

| Jan | Feb | Mar | Apr | May | Jun | Jul | Aug | Sep | Oct | Nov | Dec |

Wingspan: 15-17mm
Flight Period:
June - August
Foodplant:
Hawthorn
National Status:
Local

Approx Length (c. 8.5mm)

Flight Graph

Smaller than *Epiblema roborana* and *E. rosaecolana* with narrower forewings (this can be hard to appreciate on lone individuals so caution is advised).

Further separated from *E. rosaecolana* by the costal strigulae (**A**) appearing quite thick (i.e not as many strigulae as *E. rosaecolana*). The intensity of the pre-tornal marking and subterminal fascia (**B**) can wear and become faint so should not be used as an identification aid.

Common on chalk downs and open woodlands where hawthorn is well established throughout much of the British Isles.

Adult moths fly from dusk, later coming to light. The larvae feed in the spun shoots of the foodplant in April and May.

Similar Species: *Epiblema rosaecolana* Page **49**   *Epiblema roborana* Page **50**

Tortricidae: Olethreutinae

Epiblema rosaecolana (Doubleday, 1850) [1177]

Photo © Shane Farrell

Photo © Mike Wall (John Langmaid Collection)

| Jan | Feb | Mar | Apr | May | Jun | Jul | Aug | Sep | Oct | Nov | Dec |

Wingspan: 16-20mm
Flight Period:
June - July
Foodplant:
Wild Rose, Cultivated Rose
National Status:
Common

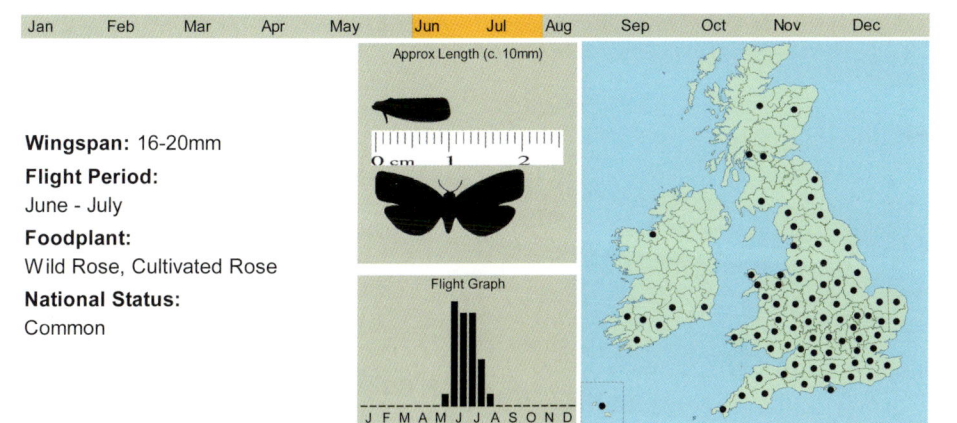

Differs from **Epiblema trimaculana** by its larger size, broader forewings and the finer, more oblique costal strigulae (**A**) (i.e the small black dashes along the leading edge of the forewing are finer and more numerous than in *E. trimaculana* and *E. roborana*). This can be difficult to view and interpret even with a good hand lens and some individuals may require genitalia examination to confirm.

Common on chalk downs, open woodlands, hedgerows and gardens throughout much of the British Isles. Widely distributed in the southern counties of England, the Midlands, and Wales. Rare in Scotland and Ireland.

Adult moths fly from dusk and come to light. The larvae feed in the shoots of the foodplant in May and June.

Similar Species: *Epiblema roborana* Page **50**   *Epiblema trimaculana* Page **48**

Tortricidae: Olethreutinae

Epiblema roborana ([Denis & Schiffermüller], 1775) [1178]

Photo © Neil Sherman (Male)

Photo © Jon Clifton (Male)

Wingspan: 18-21mm

Flight Period:
May - August

Foodplant:
Wild Rose, Cultivated Rose

National Status:
Common

Very similar to *Epiblema trimaculana* and *E. rosaecolana*, being larger than *E. trimaculana*, roughly the same size as *E. rosaecolana* (this will not be immediately apparent with lone individuals so care should be taken).

The male only differs from *E. rosaecolana* by the dark brown basal patch extending along the costal fold (**A**), which is diagnostic. Females are more problematical but should be separable by the costal strigulae being thicker and less numerous. Individual variation occurs in some populations so care is advised.

Fairly common and well distributed throughout the British Isles in hedgerows, chalk downs and open woodlands, especially where wild rose occurs. Also found on coastal cliffs and sandhills where it is sometimes plentiful on burnet rose.

Adult moths fly from dusk and come to light. The larvae feed on the leaves and flower buds of the foodplant in May and June.

Similar Species: *Epiblema trimaculana* Page **48**   *Epiblema rosaecolana* Page **49**

Tortricidae: Olethreutinae

Epiblema incarnatana (Hübner, 1800) [1179]

Photo © Helen Bantock

Photo © Mike Wall (John Langmaid Collection)

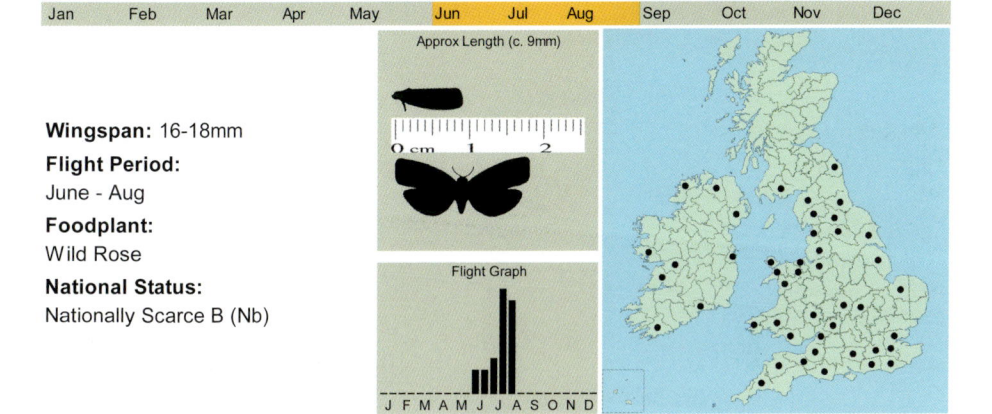

| Jan | Feb | Mar | Apr | May | Jun | Jul | Aug | Sep | Oct | Nov | Dec |

Approx Length (c. 9mm)

Wingspan: 16-18mm

Flight Period:
June - Aug

Foodplant:
Wild Rose

National Status:
Nationally Scarce B (Nb)

Flight Graph

Separated from other *Epiblema* by the delicate pink suffusion of the forewing (**A**), this can weaken with wear but should still show some ochreous coloration. Minor variation can occur with the strength of pink, if in doubt, retain the specimen.

Fairly uncommon throughout the British Isles, occurring in coastal limestone areas, sandhills, chalk downs and occasionally open woodlands. Rare in Scotland with one record from Kircudbrightshire. Scattered records in Ireland.

Adult moths fly from late afternoon and later come to light. The larvae feed in the spun leaves of the foodplant in May and June.

Similar Species: *Epiblema roborana* Page 50

Tortricidae: Olethreutinae
Epiblema tetragonana (Stephens, 1834) [1180]

Photo © Mike Wall

Photo © Ben Hossi

| Jan | Feb | Mar | Apr | May | Jun | Jul | Aug | Sep | Oct | Nov | Dec |

Wingspan: 13-16mm
Flight Period:
July
Foodplant:
Wild Rose
National Status:
Nationally Scarce B (Nb)

Superficially similar to <u>female</u> *Epiblema cirsiana/scutulana* and both *Epiblema cnicicolana/sticticana*. A reasonable amount of experience is required for confident separation of this group, if in doubt, retain the specimen for expert opinion (a photograph will not be sufficient)

E. tetragonana flies later and is generally a slightly smaller species with a shorter forewing (hard to discern in lone examples). In fresh specimens the overall blue-grey forewing ground colour and distinctive blue-grey striae (streak) arising from the costal strigulae and running parallel to the subterminal fascia (A) is visible.

A declining Nationally scarce species with very few modern records. Found in open woodlands and hedgerows at scattered locations across the British Isles.

Adult moths fly from late afternoon until dusk, later occasionally coming to light. The larvae feed in the spun leaves of the foodplant in May and June.

Tortricidae: Olethreutinae

Epiblema scutulana ([Denis & Schiffermüller], 1775) [1184]

Photo © Rob Lee

Photo © Jon Clifton (Male)

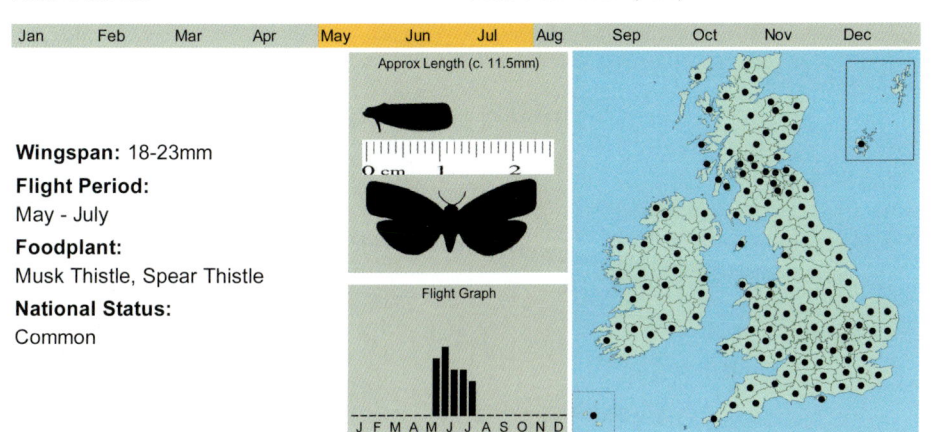

| Jan | Feb | Mar | Apr | May | Jun | Jul | Aug | Sep | Oct | Nov | Dec |

Approx Length (c. 11.5mm)

Wingspan: 18-23mm
Flight Period:
May - July
Foodplant:
Musk Thistle, Spear Thistle
National Status:
Common

Flight Graph

J F M A M J J A S O N D

Very similar to **Epiblema cirsiana**. Male *E. scutulana* can be separated from male *E. cirsiana*, on average, by its larger size and paler hindwings (A). Females of both species are very similar, *E. cirsiana* being somewhat darker overall but are problematical and would probably require genitalia examination (even this can be troublesome).

Fairly common throughout the British Isles. Found on uncultivated ground, rough meadows, hillsides, coastal cliffs and open woodlands, often preferring damp or marshy ground.

Adult moths fly actively in the evening and later come to light. The larvae feed in the roots and stems of the foodplant from August to October, over-wintering fully fed.

Similar Species: *Epiblema cirsiana* Page **54**

Tortricidae: Olethreutinae

Epiblema cirsiana (Zeller, 1843) [1184a]

Photo © Patrick Clement

Photo © Jon Clifton (Male)

Wingspan: 12-23mm
Flight Period:
May - June
Foodplant:
Marsh Thistle, Common Knapweed
National Status:
Uncertain

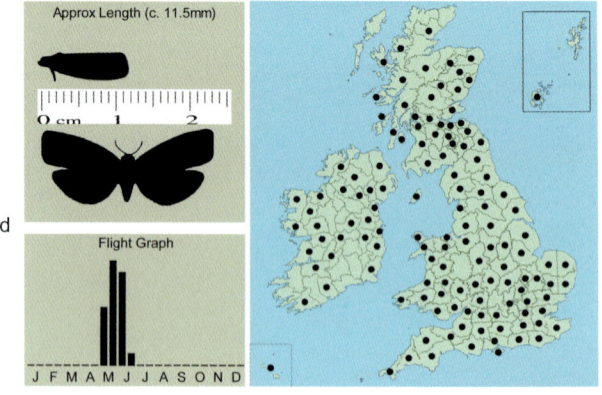

Very similar to **E. scutulana**. Male E. cirsiana can be separated from male E. scutulana by the usually smaller size and darker hindwing colour (**A**). Females of both species almost inseparable. Genitalia examination would be required.

Fairly common throughout the British Isles. The true status of this species remains uncertain owing to confusion with E. scutulana, which was formerly treated as conspecific.

Adult moths fly in the evening and come to light. The larvae feed in the roots and stems of the foodplant from August to September.

Similar Species: *Epiblema scutulana* Page 53

Tortricidae: Olethreutinae

Epiblema cnicicolana (Zeller, 1847) [1185]   **R**

Photo © James Roberts

Photo © Mike Wall (John Langmaid Collection)

| Jan | Feb | Mar | Apr | **May** | **Jun** | **Jul** | Aug | Sep | Oct | Nov | Dec |

Wingspan: 14-16mm

Flight Period:
May - June

Foodplant:
Common Fleabane

National Status:
Proposed Red Data Book (pRDB3) Category 3: Rare

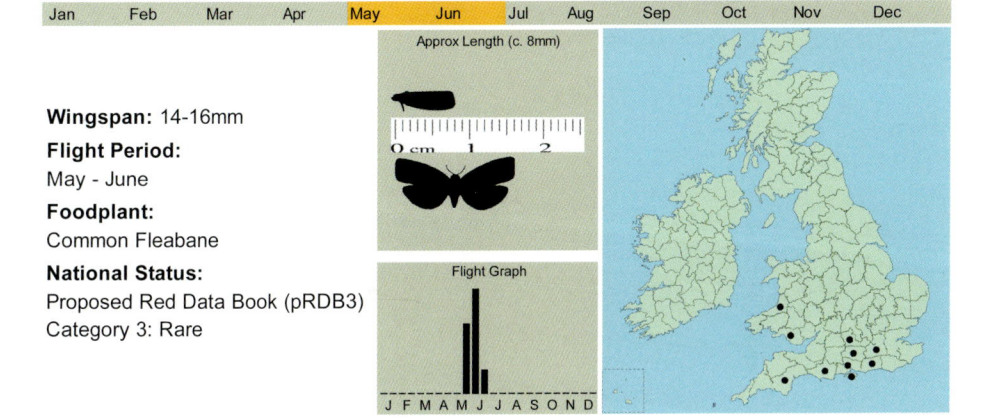

Approx Length (c. 8mm)

Flight Graph

Can resemble underlined female *Epiblema scutulana* and *E. cirsiana* and can usually be told apart by its darker brown forewing markings.

E. cnicicolana usually flies slightly earlier than *E. sticticana* and is on average, slightly smaller showing an overall darker brown forewing and slightly narrower medio-dorsal blotch (**A**), but in most cases, genitalia examination would be required

Rare, proposed Red Data Book species found in grassland, river-banks, ditches and wet meadows. Recorded from only a few counties in southern England and south Wales.

Adult moths are occasionally active in afternoon sunshine but normally early evening. The larvae feed in the lower stems of the foodplant from July to September.

Similar Species: *Epiblema scutulana* Page **53**   *Epiblema cirsiana* Page **54**

Tortricidae: Olethreutinae

Epiblema sticticana (Fabricius, 1794) [1186]
(=farfarae)

Photo © Jim Wheeler    Photo © Jim Wheeler

| Jan | Feb | Mar | Apr | May | Jun | Jul | Aug | Sep | Oct | Nov | Dec |

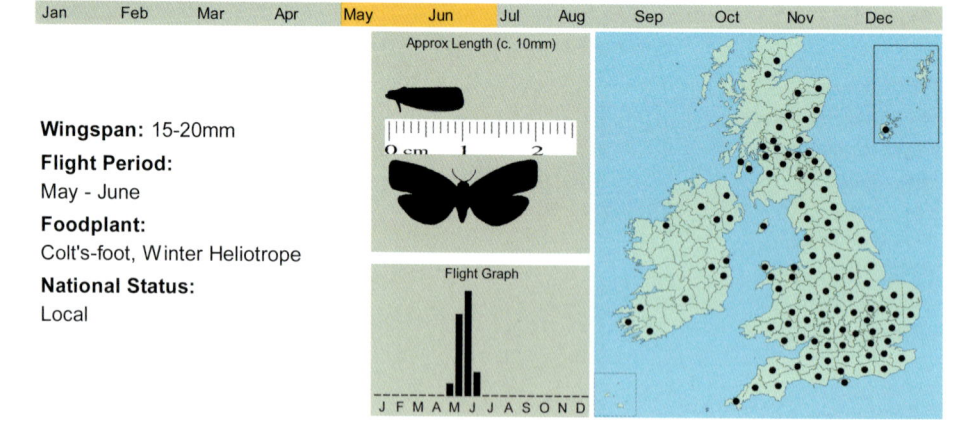

Wingspan: 15-20mm
Flight Period:
May - June
Foodplant:
Colt's-foot, Winter Heliotrope
National Status:
Local

Approx Length (c. 10mm)

Flight Graph
J F M A M J J A S O N D

Resembles **Epiblema scutulana** and **E. cirsiana** and separation is problematical but can usually be told apart by its lighter brown forewing markings.

E. sticticana is on average larger than **Epiblema cnicicolana** showing a paler olivaceous-brown forewing. Variation does occur especially in the females which are larger and show much darker brown forewings. Caution is advised and in most cases, genitalia examination would be required.

Widespread and locally common throughout the British Isles, often seen flying actively in the afternoon sunshine.

Adult moths fly actively in the afternoon sunshine, continuing until dusk. The larvae feed in the roots then flowers of the foodplant from July to April.

Similar Species: Epiblema scutulana Page 53   Epiblema cirsiana Page 54

Tortricidae: Olethreutinae

Epiblema costipunctana (Haworth, 1811) [1187]

Photo © Neil Sherman

Photo © Mark Parsons

| Jan | Feb | Mar | Apr | May | Jun | Jul | Aug | Sep | Oct | Nov | Dec |

Wingspan: 13-18mm

Flight Period:
Two Generations
May - June, July - August

Foodplant:
Common Ragwort

National Status:
Local

Approx Length (c. 9mm)

Flight Graph

Fairly small and nondescript. Distinguished from **Epiblema cnicicolana**, **E. sticticana** and **E. tetragonana,** by the richer brown basal patch and median fascia (**A**), the better marked costal strigulae (fine dashes along the costa) (**B**) and by the pale dorsal blotch showing weak strigulae (**C**).

Can often be found flying in late afternoon sunshine around the foodplant. Found on open uncultivated and waste ground, meadows, embankments, cuttings, sandhills and similar habitat throughout the British Isles.

Two generations in the south, with only one in Scotland from May to early July.

Adult moths fly actively during late afternoon and from dusk onwards, occasionally coming to light. The larvae feed in the roots and stems of the foodplant from September to December.

Tortricidae: Olethreutinae

Eucosma campoliliana ([Denis & Schiffermüller], 1775) [1197]

Photo © Allan Drewitt

Photo © Jon Clifton

Wingspan: 13-18mm

Flight Period:
June - July

Foodplant:
Common Ragwort

National Status:
Local

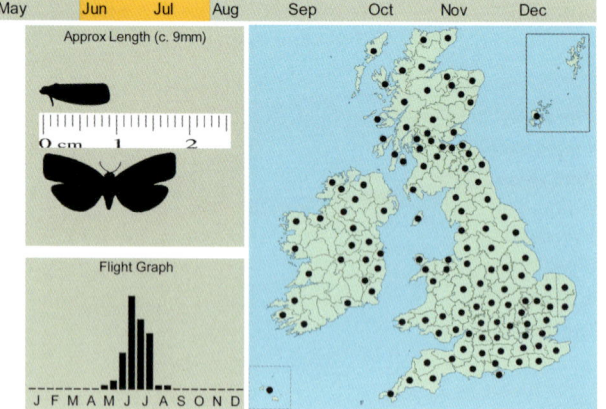

A distinctive species with white head and white ground colour. An incomplete basal patch consisting of irregular black markings (A), the median fascia being narrow on costa, thicker dorsally (B), rich brown apical spot (C).

Widely distributed, occurring in grassland, gravel pits, waysides and sand dunes, with a preference for coastal areas and well-drained, dry localities throughout the British Isles.

Adult moths fly from late evening later coming to light. The larvae feed on the seeds of the foodplant in August and September.

Tortricidae: Olethreutinae

Eucosma pauperana (Duponchel, 1843) [1198]

R

Photo © Jim Wheeler

Photo © Jon Clifton

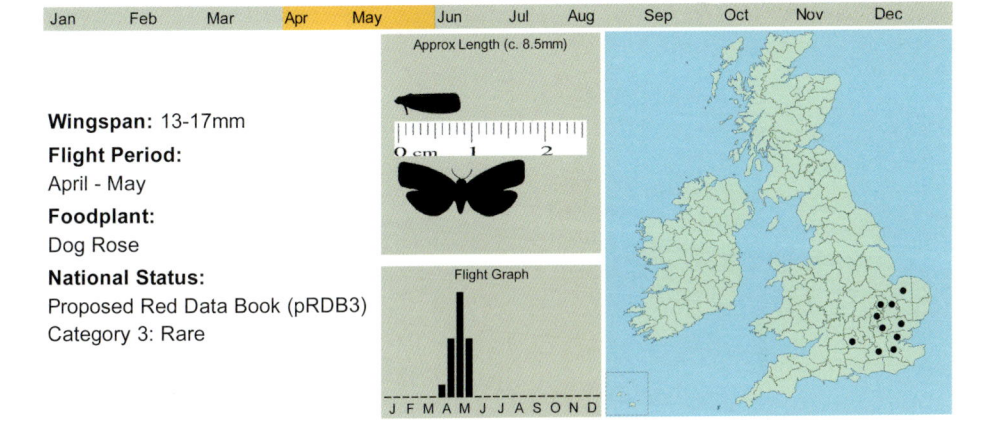

| Jan | Feb | Mar | Apr | May | Jun | Jul | Aug | Sep | Oct | Nov | Dec |

Approx Length (c. 8.5mm)

Wingspan: 13-17mm

Flight Period:
April - May

Foodplant:
Dog Rose

National Status:
Proposed Red Data Book (pRDB3) Category 3: Rare

Flight Graph

J F M A M J J A S O N D

Fairly distinctive. A large well defined basal patch, the outer edge near costa, oblique (not straight) (**A**), small black markings in pre-tornal area (**B**), the subterminal fascia showing several black raised scales (**C**).

Nationally Scarce, found on rough uncultivated ground, especially chalkland and coastal cliffs in the south eastern counties of England.

An old record from Norfolk (Fenn, 1978) noted in a paper at the time as not only a new county record, but the first specimen to be taken in Britain for 47 years. (*Ent. Rec. J. Var., 90: 245*) Recent records from Norfolk and also known from Cambridgeshire, Huntingdonshire, Bedfordshire, Hertfordshire, Essex, Kent, Surrey and Berkshire.

Adult moths can be found on warm afternoons amongst its foodplant, occasionally coming to light. The larvae feed in the spun flowers and hips of the foodplant in June.

Tortricidae: Olethreutinae

Spilonota ocellana ([Denis & Schiffermüller], 1775) [1205]
Bud Moth

Photo © Lee Gregory

Photo © Jon Clifton

Wingspan: 12-17mm

Flight Period:
July - August

Foodplant:
Various trees and shrubs

National Status:
Common

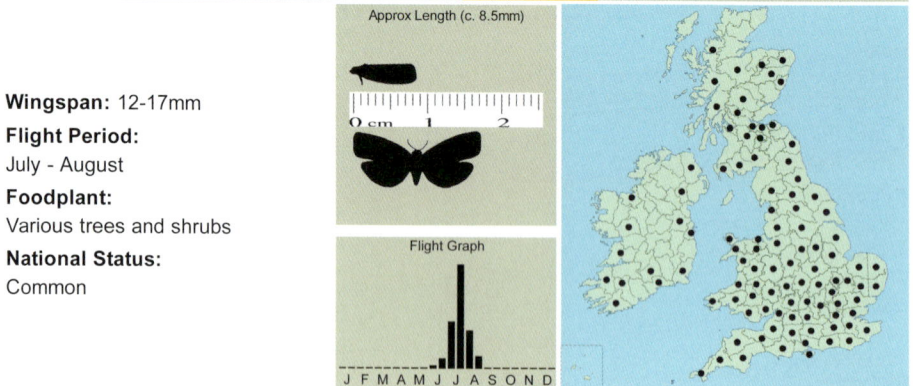

Distinguished by a well defined basal patch (**A**), and a black well developed triangular pre-tornal mark (**B**).

Could be confused with **Gypsonoma aceriana** but the larger triangular black pre-tornal mark of *S. ocellana* and the presence of several black dashes in the apical area of the ocellus (**C**) will rule this species out.

Dark forms of *S. ocellana* are widespread and could easily be confused with **Spilonota laricana**, but can be separated by the broader forewing.

Widespread in open woodlands, hedgerows, orchards and gardens. Commonly known as the *Bud Moth* because of the damage the larvae cause to the fruit buds in orchards during spring.

Adult moths fly from dusk and come to light. The larvae feed from August to June on a wide variety of plants, overwintering in a silken hibernaculum.

Similar Species: *Spilonota laricana* Page 61   *Gypsonoma aceriana* Page 42

Tortricidae: Olethreutinae

Spilonota laricana (Heinemann, 1863) [1205a]

Photo © Neil Sherman

Photo © Jon Clifton

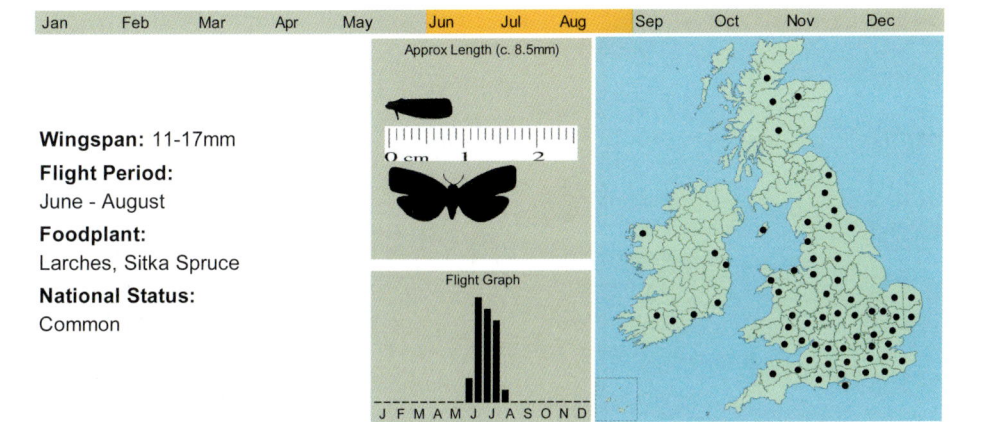

Wingspan: 11-17mm

Flight Period:
June - August

Foodplant:
Larches, Sitka Spruce

National Status:
Common

Previously thought a rather dark but distinctive form of *Spilonota ocellana* but shows a narrower forewing with the white ground colour replaced with coarse grey-black markings (**A**).

Note: the all dark form of *Spilonota ocellana* can be separated from *S. laricana* by the broader forewing. This can be hard to discern with lone examples.

Occurs mainly in the southern half of the British Isles where it can be relatively common, becoming more scattered further north.

Adult moths fly from dusk and come to light. The larvae feed in the buds and spun leaves of the foodplant in the autumn, overwintering and continuing to feed in the spring.

Similar Species: *Spilonota ocellana* Page **60**

Tortricidae: Olethreutinae

Pammene fasciana (Linnaeus, 1761) [1236]

Photo © Neil Sherman

Photo © Jon Clifton (Female)

Wingspan: 13-17mm

Flight Period:
June - August

Foodplant:
Oak, Sweet Chestnut

National Status:
Common

A characteristic species with a large pale/whitish curving dorsal blotch (**A**), several black dots set in the median fascia (**B**) almost contiguous with blackish dots in the ocellus, this being edged with lilac (**C**). Females show a darker hindwing.

Partially diurnal, the males fly soon after sunrise and again in the afternoon sunshine high up in the canopy. Occurs in woodlands, scrub, parkland, heathland, fens and gardens throughout much of England and Wales although apparently scarce in Scotland and Ireland.

The larvae feed in the fruit of the foodplants from August to October. Fallen acorns and chestnuts show very conspicuous larval exit holes.

Tortricidae: Olethreutinae

Cydia servillana (Duponchel, 1836) [1256]

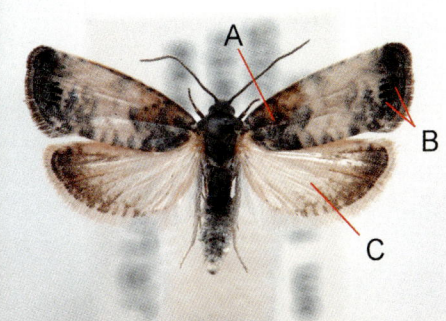

Photo © Jim Wheeler

Photo © Mike Wall (John Langmaid Collection)

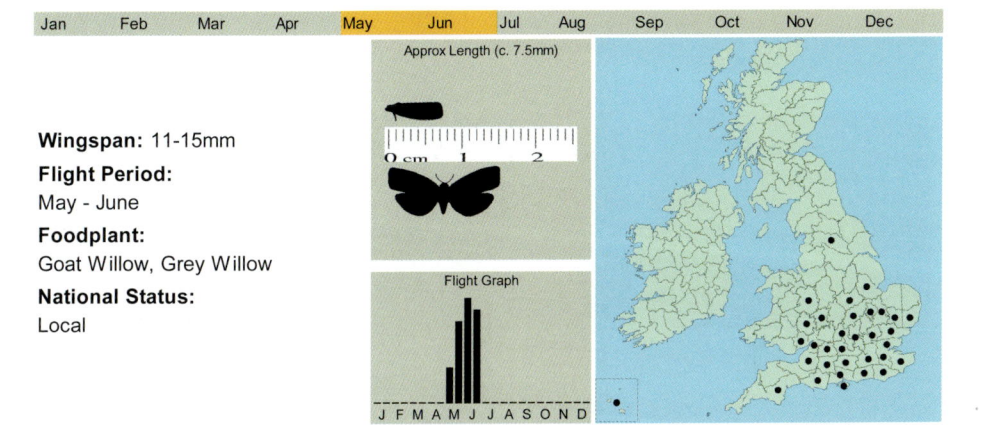

| Jan | Feb | Mar | Apr | May | Jun | Jul | Aug | Sep | Oct | Nov | Dec |

Wingspan: 11-15mm
Flight Period:
May - June
Foodplant:
Goat Willow, Grey Willow
National Status:
Local

Distinctive and should not cause confusion with any other species. The forewing ground colour being white with a hint of ochreous. Basal area suffused blue/grey and brown (**A**), ocellus conspicuous with three or four black dashes, apical area and terminal edge heavily marked with blackish/grey (**B**). Hindwing white suffused brown along the terminal edge (**C**).

Local and scarce, recorded only from southern England and south-east Wales, with an isolated record from Yorkshire. Found in damp low-lying situations often where sallow grows in hedges and ditches.

Adult males occasionally fly during afternoon and both sexes fly freely in the evening. The larvae feed in the twigs of the foodplant from August to October causing a very slender gall.

Thumb Finder

Phtheochroa sodaliana p.5

Hysterophora maculosana p.6

Phtheochroa rugosana p.7

Phalonidia manniana p.8

Eupoecilia angustana p.9

Cochylis dubitana p.10

Cochylis molliculana p.11

Cochylis hybridella p.12

Cochylis atricapitana p.13

Cochylis pallidana p.14

Cochylis nana p.15

Acleris variegana p.16

Celypha woodiana p.17

Piniphila bifasciana p.18

Hedya pruniana p.19

Hedya nubiferana p.20

Hedya ochroleucana p.21

Hedya salicella p.22

Metendothenia atropunctana p.23

Apotomis semifasciana p.24

Apotomis infida p.25

Apotomis lineana p.26

Apotomis turbidana p.27

Apotomis betuletana p.28

Apotomis capreana p.29

Apotomis sororculana p.30

Apotomis sauciana p.31

Endothenia gentianaeana p.32

Endothenia oblongana p.33

Endothenia marginana p.34

Lobesia abscisana p.35	Ancylis laetana p.36	Epinotia subocellana p.37
Epinotia bilunana p.38	Epinotia ramella p.39	Epinotia demarniana p.40
Epinotia trigonella p.41	Gypsonoma aceriana p.42	Gypsonoma sociana p.43
 Gypsonoma dealbana p.44	Gypsonoma oppressana p.45	Gibberifera simplana p.46
 Epiblema cynosbatella p.47	  Epiblema trimaculana p.48	 Epiblema rosaecolana p.49

Epiblema roborana p.50	Epiblema incarnatana p.51	Epiblema tetragonana p.52
Epiblema scutulana p.53	Epiblema cirsiana p.54	Epiblema cnicicolana p.55
Epiblema sticticana p.56	Epiblema costipunctana p.57	Eucosma campoliliana p.58
Eucosma pauperana p.59	Spilonota ocellana p.60	Spilonota laricana p.61
Pammene fasciana p.62	Cydia servillana p.63	

Photography

Allan Drewitt
Acleris variegana, Apotomis betuletana, Apotomis capreana, Apotomis lineana, Apotomis turbidana, Cochylis atricapitana, Cochylis hybridella, Epiblema cynosbatella, Epinotia bilunana, Epinotia demarniana, Epinotia ramella, Eucosma campoliliana, Eupoecilia angustana, Hedya nubiferana, Hedya pruniana, Hedya salicella, Lobesia abscisana, Phtheochroa rugosana, Phtheochroa sodaliana, Piniphila bifasciana.

Helen Bantock
Epiblema incarnatana.

Jim Wheeler
Apotomis sauciana, Apotomis semifasciana, Apotomis sororculana, Cochylis pallidana, Cydia servillana, Endothenia oblongana, Epiblema sticticana, Epiblema trimaculana, Eucosma pauperana, Gypsonoma dealbana, Gypsonoma sociana.

Mark Parsons
Celypha woodiana.

Mike Crewe
Endothenia gentianaeana, Phalonidia manniana.

Mike Wall
Cochylis molliculana, Epiblema tetragonana, Epinotia trigonella.

Neil Sherman
Ancylis laetana, Cochylis dubitana, Cochylis nana, Epiblema costipunctana, Epiblema roborana, Gypsonoma aceriana, Hysterophora maculosana, Metendothenia atropunctana, Pammene fasciana, Spilonota laricana.

Patrick Clement
Epiblema cirsiana, Gypsonoma oppressana.

Perry Hampson
Hedya ochroleucana.

Peter Buchner
Gibberifera simplana.

Rob Lee
Epiblema scutulana, Epinotia subocellana.

Shane Farrell
Epiblema rosaecolana.

Tristan Bantock
Endothenia marginana.

Set specimen photography and annotation by **Jon Clifton**.
Additional set specimen photography by **Mike Wall**, **Mark Parsons**, **Harry Beaumont** and **Graham Finch**.
All set specimens in the collection of the photographer unless otherwise noted.

References

Baixeras, J., J. W. Brown & T. M. Gilligan. 2010. T@RTS: Online World Catalogue of the Tortricidae (Version 1.4.0). http://www.tortricidae.com/catalogue.asp.
Baldwin, J. 1878. Fenland - Past and Present, Lepidoptera listed by SH Miller & SBJ Skertchly.
Barrett, C. G. 1904-05. The lepidoptera of the British Islands, 10: 1-381, pls. 425-469. London
Barrett, C. G. 1905-07. The lepidoptera of the British Islands, 11: i-lxxv, 1-293, pls. 470-504. London
Bradley, J. D. 1959. An illustrated list of the British Tortricidae. Part II: Olethreutinae. Entomologist's Gaz., 10: 60-80, text-figs. 1-12, pls. 1-19
Bradley, J.D. Checklist of Lepidoptera Recorded from The British Isles, Second Edition (Revised) (2000)
Bradley, J.D., Tremewan, W.G. and Smith, A. 1973. British Tortricoid Moths. Cochylidae and Tortricidae: Tortricinae. The Ray Society.
Bradley, J.D., Tremewan, W.G. and Smith, A. 1979. British Tortricoid Moths. Tortricidae: Olethreutinae. The Ray Society.
Emmet, A. M. (ed). 1979. A Field Guide to the smaller British Lepidoptera. – The British Entomological and Natural History Society, London. 271 pp.
Ferguson, I.D. 2003. Checklist of Kent Lepidoptera.
Fenn, J. 1978. Ent. Rec. J. Var., 90: 245
Kloet, G.S. & Hincks, W.D. 1976, A check list of British Insects. Second edition (completely revised). Part 2: Lepidoptera.
Manley, C. 2008. British Moths and Butterflies: A Photographic Guide. A & C Black.
Razowski, Jozef. 2001. Die Tortriciden (Lepidoptera, Tortricidae) Mitteleuropas : Bestimmung, Verbreitung, Flugstandort, Lebensweise der Raupen. Bratislava.
Razowski, Jozef. 2002. Tortricidae of Europe. Part I. Tortricinae and Chlidanotinae. Frantisek Slamka, Bratislava.
Razowski, Jozef. 2003. Tortricidae of Europe. Part II. Olethreutinae. Frantisek Slamka, Bratislava.
Szaboky, C., Csoka, G. 2010. Tortricids / Sodromolyok. Forest Research Institute.
Whittle, F.G. 1920. Lepidoptera at Rannoch in 1919. Entomologist, 53: 11 - 13

Websites

Hants Moths - www.hantsmoths.org.uk
Lepiforum - www.lepiforum.de
NBN Gateway - www.nbn.org.uk
Norfolk Moths - www.norfolkmoths.org.uk
Suffolk Moths - www.suffolkmothgroup.org.uk
UK Moths - www.ukmoths.org.uk

Species Index

Acleris variegana (1048) Page **16**
Ancylis laetana (1123) Page **36**
Apotomis betuletana (1093) Page **28**
Apotomis capreana (1094) Page **29**
Apotomis infida (1090) Page **25**
Apotomis lineana (1091) Page **26**
Apotomis sauciana (1096) Page **31**
Apotomis semifasciana (1089) Page **24**
Apotomis sororculana (1095) Page **30**
Apotomis turbidana (1092) Page **27**
Celypha woodiana (1066) Page **17**
Cochylis atricapitana (0966) Page **13**
Cochylis dubitana (0964) Page **10**
Cochylis hybridella (0965) Page **12**
Cochylis molliculana (0964a) Page **11**
Cochylis nana (0968) Page **15**
Cochylis pallidana (0967) Page **14**
Cydia servillana (1256) Page **63**
Endothenia gentianaeana (1097) Page **32**
Endothenia marginana (1099) Page **34**
Endothenia oblongana (1098) Page **33**
Epiblema cirsiana (1184a) Page **54**
Epiblema cnicicolana (1185) Page **55**
Epiblema costipunctana (1187) Page **57**
Epiblema cynosbatella (1174) Page **47**
Epiblema incarnatana (1179) Page **51**
Epiblema roborana (1178) Page **50**
Epiblema rosaecolana (1177) Page **49**
Epiblema scutulana (1184) Page **53**
Epiblema sticticana (1186) Page **56**
Epiblema tetragonana (1180) Page **52**
Epiblema trimaculana (1176) Page **48**
Epinotia bilunana (1133) Page **38**
Epinotia demarniana (1135) Page **40**
Epinotia ramella (1134) Page **39**
Epinotia subocellana (1132) Page **37**
Epinotia trigonella (1151) Page **41**
Eucosma campoliliana (1197) Page **58**
Eucosma pauperana (1198) Page **59**
Eupoecilia angustana (0954) Page **9**
Gibberifera simplana (1173) Page **46**
Gypsonoma aceriana (1167) Page **42**
Gypsonoma dealbana (1169) Page **44**
Gypsonoma oppressana (1170) Page **45**
Gypsonoma sociana (1168) Page **43**
Hedya nubiferana (1083) Page **20**
Hedya ochroleucana (1084) Page **21**
Hedya pruniana (1082) Page **19**
Hedya salicella (1086) Page **22**
Hysterophora maculosana (0924) Page **6**
Lobesia abscisana (1108) Page **35**
Metendothenia atropunctana (1085) Page **23**
Pammene fasciana (1236) Page **62**
Phalonidia manniana (0926) Page **8**
Phtheochroa rugosana (0925) Page **7**
Phtheochroa sodaliana (0923) Page **5**
Piniphila bifasciana (1079) Page **18**
Spilonota laricana (1205a) Page **61**
Spilonota ocellana (1205) Page **60**

Back Cover Photographs

Epiblema trimaculana by **Jim Wheeler**
Metendothenia atropunctana by **Neil Sherman**
Pammene fasciana by **Neil Sherman**

Design & Layout by Jim Wheeler